Frauke Mager

**Structural Insights into the Proteintranslocase TOM**

Frauke Mager

# Structural Insights into the Proteintranslocase TOM

The translocase of the outer mitochondrial membrane from Neurospora crassa and Homo sapiens

Südwestdeutscher Verlag für Hochschulschriften

**Impressum/Imprint (nur für Deutschland/only for Germany)**
Bibliografische Information der Deutschen Nationalbibliothek: Die Deutsche Nationalbibliothek verzeichnet diese Publikation in der Deutschen Nationalbibliografie; detaillierte bibliografische Daten sind im Internet über http://dnb.d-nb.de abrufbar.
Alle in diesem Buch genannten Marken und Produktnamen unterliegen warenzeichen-, marken- oder patentrechtlichem Schutz bzw. sind Warenzeichen oder eingetragene Warenzeichen der jeweiligen Inhaber. Die Wiedergabe von Marken, Produktnamen, Gebrauchsnamen, Handelsnamen, Warenbezeichnungen u.s.w. in diesem Werk berechtigt auch ohne besondere Kennzeichnung nicht zu der Annahme, dass solche Namen im Sinne der Warenzeichen- und Markenschutzgesetzgebung als frei zu betrachten wären und daher von jedermann benutzt werden dürften.

Coverbild: www.ingimage.com

Verlag: Südwestdeutscher Verlag für Hochschulschriften GmbH & Co. KG
Dudweiler Landstr. 99, 66123 Saarbrücken, Deutschland
Telefon +49 681 37 20 271-1, Telefax +49 681 37 20 271-0
Email: info@svh-verlag.de

Approved by: Stuttgart, Universität, Diss., 2011

Herstellung in Deutschland:
Schaltungsdienst Lange o.H.G., Berlin
Books on Demand GmbH, Norderstedt
Reha GmbH, Saarbrücken
Amazon Distribution GmbH, Leipzig
**ISBN: 978-3-8381-2610-4**

**Imprint (only for USA, GB)**
Bibliographic information published by the Deutsche Nationalbibliothek: The Deutsche Nationalbibliothek lists this publication in the Deutsche Nationalbibliografie; detailed bibliographic data are available in the Internet at http://dnb.d-nb.de.
Any brand names and product names mentioned in this book are subject to trademark, brand or patent protection and are trademarks or registered trademarks of their respective holders. The use of brand names, product names, common names, trade names, product descriptions etc. even without a particular marking in this works is in no way to be construed to mean that such names may be regarded as unrestricted in respect of trademark and brand protection legislation and could thus be used by anyone.

Cover image: www.ingimage.com

Publisher: Südwestdeutscher Verlag für Hochschulschriften GmbH & Co. KG
Dudweiler Landstr. 99, 66123 Saarbrücken, Germany
Phone +49 681 37 20 271-1, Fax +49 681 37 20 271-0
Email: info@svh-verlag.de

Printed in the U.S.A.
Printed in the U.K. by (see last page)
**ISBN: 978-3-8381-2610-4**

Copyright © 2011 by the author and Südwestdeutscher Verlag für Hochschulschriften GmbH & Co. KG and licensors
All rights reserved. Saarbrücken 2011

# Contents

Acknowledgements ................................................................................................ 5
Abbreviations ........................................................................................................ 6
Abstract ................................................................................................................. 9
Zusammenfassung ............................................................................................... 11
1   Introduction .................................................................................................... 13
    1.1   Mitochondria and their origin ................................................................. 13
    1.2   Protein transport into mitochondria ....................................................... 15
        1.2.1   General import mechanism ............................................................ 15
        1.2.2   Protein transport across and assembly into the outer membrane ........ 18
        1.2.3   Protein transport across and assembly into the inner membrane ......... 20
        1.2.4   Protein transport into the inner membrane space ........................... 22
    1.3   TOM Complex: function and components ............................................... 22
        1.3.1   Tom70 ............................................................................................ 24
        1.3.2   Tom20 ............................................................................................ 25
        1.3.3   Tom22 ............................................................................................ 25
        1.3.4   Tom40 ............................................................................................ 26
        1.3.5   Tom5, Tom6 and Tom7 .................................................................. 27
    1.4   Beta-barrel membrane proteins .............................................................. 27
    1.5   Aim of this study ..................................................................................... 31
2   Materials and methods ................................................................................... 32
    2.1   Equipment ............................................................................................... 32
        2.1.1   Chemicals ....................................................................................... 32
        2.1.2   Devices ........................................................................................... 32
    2.2   Microbiological methods ......................................................................... 33
        2.2.1   Bacterial cell culture ...................................................................... 33
        2.2.2   Preparation of chemically competent *E. coli* cells for transformation .... 34
        2.2.3   Transformation .............................................................................. 34
        2.2.4   Glycerol stocks ............................................................................... 35
        2.2.5   Isolation of inclusion bodies .......................................................... 35
        2.2.6   TOM complex isolation from *N. crassa* mitochondria ................. 36

2.3 Molecular biology methods .................................................................................................. 37
    2.3.1 Strains and plasmids ..................................................................................................... 37
    2.3.2 Isolation of plasmid DNA .............................................................................................. 38
    2.3.3 Agarose gel electrophoresis ......................................................................................... 38
    2.3.4 DNA extraction from agarose gels ............................................................................... 39
    2.3.5 DNA Sequencing .......................................................................................................... 39
    2.3.6 Digestion of plasmid DNA ............................................................................................ 39
    2.3.7 Ligation ........................................................................................................................ 39
    2.3.8 PCR and site-directed mutagenesis ............................................................................. 40
2.4 Biochemical Methods .......................................................................................................... 41
    2.4.1 Determination of protein concentration ..................................................................... 41
    2.4.2 SDS-PAGE ..................................................................................................................... 41
    2.4.3 Immunoblotting of proteins ........................................................................................ 45
    2.4.4 Cross-Linking ................................................................................................................ 46
    2.4.5 Protein chromatography ............................................................................................. 47
    2.4.6 Stripping and recharging of Ni-Sepharose HisTrap columns ....................................... 49
    2.4.7 Refolding screen .......................................................................................................... 49
    2.4.8 Concentration and dialysis of protein samples ........................................................... 50
2.5 Biophysical and structural methods .................................................................................... 51
    2.5.1 Dynamic light scattering .............................................................................................. 51
    2.5.2 CD spectroscopy .......................................................................................................... 51
    2.5.3 Fourier Transformation Infrared Resonance Spectroscopy (FTIR) .............................. 52
    2.5.4 Electron microscopy .................................................................................................... 53
    2.5.5 Fluorescence spectroscopy ......................................................................................... 53
    2.5.6 Laser Induced Liquid Bead Ion Desorption (LILBID) .................................................... 54
    2.5.7 Electrophysiology ........................................................................................................ 55
    2.5.8 3D-Crystallization ........................................................................................................ 58
3 Results ......................................................................................................................................... 62
    3.1 Stoichiometry of TOM core complex ................................................................................ 62
        3.1.1 Purification of TOM core complex ............................................................................ 62
        3.1.2 LILBID-mass spectrometry ........................................................................................ 64
        3.1.3 Structural characterization of NcTom40 .................................................................. 69
    3.2 Human Tom40 .................................................................................................................. 73

3.2.1 Expression, purification and refolding of human Tom40 .......... 74
3.2.2 Structural characterization .......... 77
3.2.3 Structure modelling of human Tom40 .......... 82
3.3 Stability of β-barrel membrane proteins .......... 86
3.3.1 Cloning, expression and refolding of mutated hTom40AΔ1-82 .......... 89
3.3.2 Secondary structure determination of hTom40AΔ1-82$^{3mut}$ .......... 90
3.3.3 Heat stability of hTom40AΔ1-82$^{3mut}$ .......... 92
3.3.4 Stability of hTom40AΔ1-82$^{3mut}$ in chaotropic reagents .......... 92
3.3.5 Crosslinking of wild type and mutant hTom40AΔ1-82 .......... 94
3.3.6 3D crystallization of wild type and mutant hTom40AΔ1-82 .......... 96
4 Discussion .......... 104
4.1 Mass of the TOM core complex from *N. crassa* .......... 104
4.2 Stoichiometry of the TOM complex .......... 106
4.3 Structure of TOM core complex .......... 108
4.4 Recombinant expression of different Tom40 isoforms .......... 111
4.5 Secondary structure determination: applications and limitations .......... 113
4.6 Physiological properties of Tom40 .......... 114
4.7 Protein aggregation: True or not? .......... 114
4.8 Stabilization of the β-barrel .......... 116
4.9 Crystallization .......... 117
4.10 Outlook .......... 118
5 Bibliography .......... 119
6 Appendix .......... 128

# Acknowledgements

First, I would like to thank my supervisor, Professor Stephan Nußberger, for giving me the opportunity to work in his lab, for providing me this interesting project, and for his continuous support in this research project.

Equal thanks go to Dr. Kornelius Zeth for funding me through most of my working time in the lab, for helpful discussions, and his qualified support.

I would like to thank all my coworkers in the biophysical department in Stuttgart, our exchange students as well as all other students who contributed to this work. Special thanks go to the department I at the Max Planck Institute in Tübingen for fruitful discussions and suggestions. Also, I would like to thank all collaborators that worked with me on the project.

I thank my parents for giving me the opportunity to study my preferred subject and for supporting me ever since.

Finally, I would like to thank my husband Christoph for encouraging me at every step in this sometimes difficult project, for his help with technical problems, and especially his patience and his love.

# Abbreviations

| | |
|---|---|
| AP | alkaline phosphatase |
| APS | ammonium peroxidsulfate |
| ATP | adenosine 5´-triphosphate |
| BSA | bovine serum albumin |
| CD | circular dichroism |
| CHAPS | 3-[(3-cholamidopropyl-)dimethylammonio]-1-propanesulfonate |
| C-terminus | carboxy terminus |
| β-OG | n-octyl-β-glucopyranoside |
| Δ | delta, truncation |
| Δψ | difference in the transmembrane potential |
| DDM | n-dodecyl-β-maltoside |
| DiphPC | diphtanoyl-phosphatidyl choline |
| DMPC | dimyristoyl phosphatidyl choline |
| DMSO | dimethylsulfoxide |
| dNTP | desoxyribonucleotides |
| ε | extinction coefficient |
| EDTA | ethylenediaminetetraacetic acid |
| FTIR | Fourier transformation infrared |
| g | standard gravity |
| GIP | general import pore |
| HEPES | 4-(2-hydroxyethyl)-1-piperazineethanesulfonic acid |
| Hsp | heat shock protein |
| HPLC | high performance/pressure liquid chromatography |
| IMS | inter membrane space |
| IPTG | isopropyl β-D-1-thiogalactopyranoside |
| IR | infrared |
| (k)Da | (kilo)Dalton |
| (k)Hz | (kilo)Hertz |
| LB-medium | lysogeny broth |
| LDAO | lauryldimethylamine-oxide |
| MIM | mitochondrial inner membrane |
| MOM | mitochondrial outer membrane |
| Ni-NTA | nickel-nitrilotriacetic acid |
| NMR | nuclear magnetic resonance |
| N-terminus | amino terminus |
| OD | optical density |

| | |
|---|---|
| OMV | outer membrane vesicles |
| oPOE | n-octyl-polyoxyethylene |
| PBS | phosphate buffered saline |
| PCR | polymerase chain reaction |
| PEG | polyethylenglycol |
| (p)F | (piko) Farad |
| PMSF | phenylmethylsulfonylfluoride |
| PVDF | polyvinyliden-difluoride |
| rpm | rounds per minute |
| SAXS | small angle x-ray scattering |
| SEC | Size exclusion chromatography |
| SDS | sodiumdodecylsulfate |
| SDS-PAGE | sodiumdodecylsulfate polyacrylamide gelelectrophoresis |
| TEMED | N,N,N´,N´-tetramethylene diamine |
| Tris-HCl | tris-(hydroxymethyl)-aminoethane hydrochloric acid |
| U | Unit(s) |
| UV | ultra violet |
| VDAC | voltage-dependent anion channel |
| Vol. | Volume |
| v/v | volume per volume |
| wt | wild type |
| w/v | weight per volume |

# Abstract

The translocase of the outer mitochondrial membrane TOM is responsible for the transport of all nuclear-encoded proteins into mitochondria. The TOM complex consists of several subunits and their composition in fungi, mammals and plants is remarkably similar. Although the subunit composition of the TOM complex is known, its stoichiometry is still a matter of controversy. In this study, the subunit composition, mass, and stoichiometry of purified *Neurospora crassa* TOM core complex was determined with laser-induced ion desorption coupled to mass spectrometry. The results gave hints about the mode and strength of interaction between single subunits in the TOM complex.

The main constituent of the mitochondrial protein translocase TOM is the channel-forming protein Tom40. It belongs to the mitochondrial porin family and represents the only essential subunit of the high molecular mass TOM complex. This study describes the recombinant expression, purification, and folding of two human Tom40 isoforms for structural biology experiments. Secondary structure analyses revealed a dominant β-sheet structure and a small α-helical content in connection with a high thermal stability for both proteins. Channel activity measurements with both Tom40 isoforms in planar lipid bilayers confirmed their functionality as pore proteins.

The β-strands in membrane proteins contribute to an individual degree to the overall stability of the protein fold. To increase the stability of Tom40 for crystallographic studies, the potential energetic contribution of the predicted β-strands was calculated using bioinformatics tools. In human Tom40, three rather unstable β-strands in the transmembrane domain were detected in this study. To examine the destabilizing effects of these strands, key amino acids in each of the three strands were substituted by hydrophobic amino acids using site-directed mutagenesis. Thermal stability and solvent denaturation were examined and revealed a significant stabilization of the mutant Tom40. The tendency to oligomerize, which may shield unstable β-strands, was reduced in the mutant protein. The improved stability of the mutant Tom40 provides a base for crystallographic studies in the future.

# Zusammenfassung

Die Translokase der äußeren Mitochondrienmembran TOM ist verantwortlich für den Transport aller kerncodierter Proteine in Mitochondrien. Der TOM-Komplex besteht aus mehreren Untereinheiten, deren Zusammensetzung in Pilzen, Säugern und Pflanzen ausgesprochen ähnlich ist. Obwohl die Anzahl der Untereinheiten im TOM-Komplex bekannt ist, wird ihre Stöchiometrie weiter intensiv diskutiert. In dieser Arbeit wurde die Masse, Zusammensetzung und Stöchiometrie des gereinigten TOM-Komplexes aus *Neurospora crassa* mit laserinduzierter Ionendesorption, gekoppelt an Massenspektrometrie, analysiert. Außerdem gaben die Ergebnisse Hinweise über die Art und Stärke der Interaktionen zwischen den einzelnen Untereinheiten und deren Organisation im TOM-Komplex.

Die Hauptuntereinheit der mitochondrialen Proteintranslokase TOM ist das Kanalprotein Tom40. Es gehört zur Familie der mitochondrialen Porine und ist die einzige essentielle Untereinheit des hochmolekularen TOM-Komplexes. Diese Arbeit beschreibt die rekombinante Expression, Reinigung und Rückfaltung der zwei humanen Tom40-Isoformen für strukturbiologische Untersuchungen. Sekundärstrukturbestimmungen zeigten für beide Proteine eine ausgeprägte β-Faltblatt Struktur und einen kleinen α-helikalen Anteil, verbunden mit einer hohen thermischen Stabilität. Die Kanalaktivität von rekombinantem Tom40 in planaren Doppellipidmembranen bestätigte die native Funktion als Porenprotein.

Die transmembranen β-Stränge in Membranproteinen tragen in unterschiedlicher Weise zur gesamten Stabilität der Proteinfaltung bei. Um die Stabilität von Tom40 für röntgenkristallographische Studien zu erhöhen, wurden die Energielevel für alle vorhergesagten β-Stränge des Proteins berechnet. Dabei wurden drei instabile Stränge in humanem Tom40 identifiziert, wofür jeweils eine Aminosäure pro Strang verantwortlich war. Um den destabilisierenden Effekt dieser Stränge zu analysieren, wurden diese drei Aminosäuren mittels gerichteter Mutagenese durch hydrophobe Aminosäuren ausgetauscht. Thermische Analyse und Faltungsverhalten in chaotropen Reagentien zeigten eine signifikante Stabilisierung des mutierten Proteins. Die Oligomerisierung des Proteins, durch die instabile Stränge von der Umgebung abgeschirmt werden können, war im mutierten Tom40 reduziert. Die verbesserte Stabilität des mutierten Proteins stellt eine Grundlage für die kristallographische Strukturbestimmung von Tom40 dar.

# 1 Introduction

## 1.1 Mitochondria and their origin

The formation of an enclosed lipid vesicle in the cell was the first step on the way to cell organelles. Hence, new reaction centers were established where the enclosed environment allowed chemical processes in a restricted area. An allocation of functions to organelles is a fundamental step towards high evolved cells. Surrounded by a lipid membrane they allow for the separation of different reaction centers and autonomous organization of metabolic pathways (Voet 1994). Biological membranes surrounding these organelles mainly consist of phospholipids and act as semipermeable barriers to ions and macromolecules. The permeability of such membranes is mainly determined by their inventory of membrane-integrated proteins. While receptors, transporters and ion channels have high substrate specificity porins are responsible for gradient-driven diffusion of rather low selectivity. The membrane components concerning lipid and protein composition differ among organisms and organelles and represent key factors for membrane structure and stability and the function of the respective organelle.

The first double-walled cell organelle was probably established by the uptake of a $\alpha$-proteo-bacterium into a eukaryotic precursor cell (Sagan 1967). Due to the similar arrangement of genes on the mitochondrial genome among species it is believed that all double-walled cell organelles descent from one single endosymbiontic uptake (Gray 1999). Starting from this common eukaryotic ancestor to the complex extant organism mitochondria evolved independently over a time range of approximately 1.5 billion years. For example, the human pathogen *Giardia intestinalis* comprises a double-walled precursor organelle or reduced mitochondrion. This so-called "mitosome" of *G. intestinalis* is already in charge of energy production (Dolezal et al. 2005; Dagley et al. 2009) and symbolizes a link on the way to mitochondria formation in eukaryotes.

Today, mitochondria are a unique feature of eukaryotic cells. Their size and shape differs in respect of cell type but all exhibit two membranes. The mitochondrial outer membrane (MOM) has a structure comparable to the cytosolic membrane of bacteria while the inner membrane (MIM) has invaginations, called cristae, and therefore a much larger surface (Palade 1952). Mitochondria contain two aqueous compartments, the inter-membrane-space (IMS) and the matrix. The protein composition in both membranes differs widely and therefore comprises different functions. A main aspect of mitochondria is the energy production for the host cell which is taking place in the inner mitochondrial membrane. The production of ATP in accordance to the aerobe cell oxidation marks the main duty of mitochondria. Pyruvate coming from the glycolysis in the cell cytosol gets oxidized in the mitochondrial matrix in the citrate cycle, and its products get transformed to ATP in the respiratory chain whose components sit in the inner membrane. As protons move down the electrochemical gradient, ATP gets synthesized by the $F_0/F_1$ ATP Synthase in the inner membrane. This reaction is driven by the conversion of Gibbs free energy derived from trans-membrane electrochemical proton gradient over the inner membrane (Mitchell 1966). This requires a tight sealing of the inner membrane. The outer membrane, in contrast, is permeable to small molecules like ions and nucleotides which diffuse freely through the voltage dependent anion channel (VDAC) (Colombini 1979; Rostovtseva and Bezrukov 1998).

Due to their heritage from the incorporation of another organism, mitochondria still contain their own genome even though it has been reduced drastically during evolution. More than 95 % of proteins needed in mitochondria are encoded in the nucleus. During evolution a gene transfer from mitochondria to the nucleus took place known as "endosymbiontic gene transfer" (Timmis et al. 2004). Concerning mitochondrial gene composition the bacterium *Holospora obtusa* has most similarities with the present mitochondrial genome (Lang 2005) which makes it a close relative to the endosymbiont once incorporated. The most abundant mitochondrial genome today can be found in *Reclinomonas americana* and includes 62 genes while human mitochondria contain only 13 protein-encoding genes (Gray et al. 1999). To ensure the complete functionality of mitochondria it is necessary to provide constant protein transport from the nucleus to the organelle.

Mitochondria also play an active role in the induction of apoptosis (Green and Reed 1998). Regulation of apoptosis is mainly controlled by the Bcl2 protein family to which the pro-apoptotic protein Bax belongs. The translocase of the outer mitochondrial membrane (TOM) has been considered as possible Bax receptor candidate (Ott et al. 2007; Colin et al. 2009). By oligomerization and pore formation Bax induces cytochrome C release into the cytosol through permeabilization of the MOM presumably also by oligomerization of VDAC (Eskes et al. 1998; Keinan et al. 2010). Cytochrome C interacts with cytosolic Apaf-1 and forms the so called apoptosome which in turn is able to bind and activate cysteine proteases called caspases. After caspase activation, a signal cascade is initiated in the cytoplasm and eventually leads to further execution of apoptosis and eventually subsequently to the degradation of cytosolic proteins and therefore the self-digestion of the cell.

## 1.2 Protein transport into mitochondria

The translocation of proteins across biological membranes is an essential process that occurs in all living organisms. Prominent examples are the transport of proteins across membranes of eukaryotic organelles, such as the endoplasmic reticulum, peroxisomes, the double-walled chloroplasts and mitochondria, and the protein secretory pathways of bacteria.

### 1.2.1 General import mechanism

As mentioned, mitochondrial genomes only encode for a small subset of their essential proteins while a vast majority of mitochondrial proteins are encoded in the nucleus. In the past 30 years considerable insight has been gained on the translocation process as numerous genes involved in protein transport into mitochondria have been identified in the budding yeast *Saccharomyces cerevisiae*, the filamentous fungus *Neurospora crassa*, plants and animals (Ramage et al. 1993; Neupert 1997; Lithgow 2000). Especially studies with yeast and *N. crassa* revealed details of the mitochondrial protein import. Biochemical and genetic studies have shed light on the molecular properties and functions of numerous complexes and their sorting of proteins into different mitochondrial sub-compartments. The transport of proteins to the mitochondrial matrix is one of the best characterized sorting processes in the cell (for review see (Neupert and Herrmann 2007; Bolender et al. 2008).

Generally, protein transport itself can be divided into two mechanisms: co-translational transport which requires tight interaction of translation and translocation of proteins and post-translational transport where proteins synthesis is not directly linked to protein translocation and requires the aid of chaperones. For the co-translational import cytosolic factors like the signal-recognition particle (SRP) are in charge to guide the ribosome to the target organelle which is the case for the protein transport into the endoplasmic reticulum. Similar "SRP"-like proteins in the mitochondrial matrix are in charge of guiding mitochondrial ribosomes to protein translocases of the inner mitochondrial membrane for insertion of mitochondrial-encoded proteins into the MIM (Jia et al. 2003). But generally, protein transport from the nucleus to mitochondria occurs post-translationally. The post-translational protein transport requires a strong cooperation of nascent peptide chains with cytosolic chaperones which hinder the preproteins from premature folding and guide them to their target organelle. The transport of proteins targeting multiple destinations is regulated by the concentration of chaperones in the cytosol (Komiya et al. 1996; Sass et al. 2003). They also protect precursors from degradation by cytosolic proteases during movement through TOM (Esaki et al. 2003; Yano et al. 2004).

Altogether, various high molecular mass complexes, such as TIM23, TIM22, TOM, SAM and Oxa, coordinate the import of about 1000 (yeast) to 1500 (human) different pre-proteins into mitochondria while only few mitochondrial proteins are synthesized in the matrix. All known protein complexes and their pathways involved in mitochondrial transport are summarized in Figure 1.1 and Table 1 and will be presented below in detail.

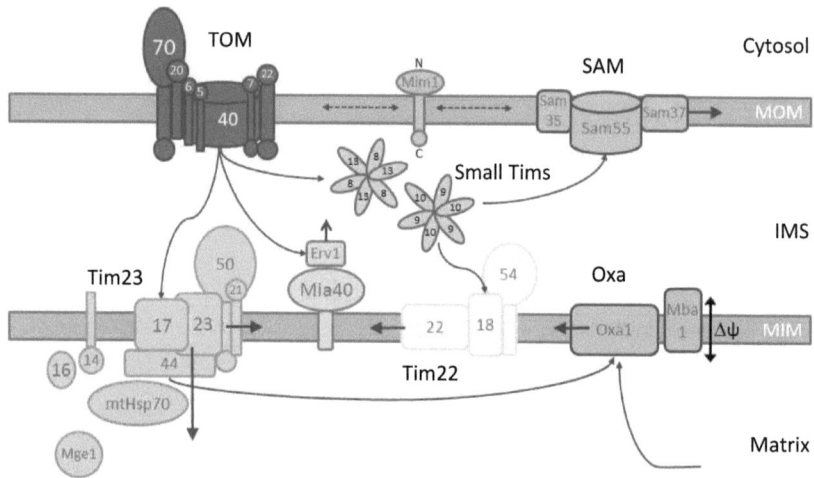

Figure 1.1: Protein translocation across the outer and inner mitochondrial membrane. Preproteins inserted by the TOM complex are transferred to the TIM23 complex and the Mia40 complex in the inner membrane. Proteins destined to the outer membrane are transported via TOM and the small Tim proteins to the SAM complex. Mim1 interacts with the SAM complex. The small Tim proteins also transport preproteins to the TIM22-complex for inner membrane insertion. The Oxa1 complex is responsible for the inner membrane insertion of proteins from the matrix or TIM23-assisted import. MOM: mitochondrial outer membrane, IMS: inner membrane space, MIM: mitochondrial inner membrane (Figure adapted from (Mokranjac and Neupert 2009; Dimmer and Rapaport 2010).

A large amount of target signals destines preproteins to the different mitochondrial compartments, the outer membrane, the intermembrane space, the inner membrane or the matrix. This sorting signal can be located at different positions in the protein (Chacinska et al. 2009; Schleiff and Becker 2010). The classical mitochondrial targeting signal guiding proteins to the matrix consists of 15-70 amino acids which are predicted to form an amphiphatic α-helix in their N-terminus. They are arginine-rich, which also represents a target for processing peptidases in the matrix which cleave the signal sequence right after import (Huang et al. 2009). Target signals of certain inner membrane proteins are located close to the hydrophobic region and some are enriched with cysteines. They require a certain length of ~ 80 amino acids to span the distance through TOM, the IMS and TIM (Matouschek et al. 1997). An example for an inner membrane protein is the ADP-ATP carrier which has a cryptic targeting signal recognized by TIM23 and is finally inserted into the inner membrane by the TIM22 complex (Vergnolle et al. 2005). The signal peptides of inner membrane proteins which are recognized by the Oxa complex comprise negatively charged side chains to become attracted by the positively charged environment in the IMS (Preuss et al. 2005). Proteins targeted to the intermembrane space passing the disulfide relay system Mia40 comprise a 9-amino-acid-signal peptide close to the processable cysteine (Sideris et al. 2009). Proteins bound for the outer membrane are translocated through the TOM complex and then handed to the SAM complex which inserts the protein in the outer membrane. Many outer membrane proteins, like mitochondrial β-barrels, share a polytopic structure in their internal sorting signals. Their correct insertion requires a conserved motif in the last β-strand, therefore termed "β-signal", which is not cleaved (Kutik et al. 2008). Nevertheless, the exact insertion mechanism of β-barrel proteins into the outer mitochondrial membrane remains poorly understood.

Table 1: Protein composition of complexes of outer and inner membrane from *N. crassa* and human

| | | *N. crassa* | Human |
|---|---|---|---|
| MOM | TOM | Tom40, Tom22, Tom20, Tom70, Tom5, 6, 7 | Tom40A and B, Tom22, Tom20, Tom70, Tom7, Tom5,6[1], Tom34 |
| | SAM | Sam50, Sam35, Sam37, Mdm10, | Sam50, Metaxin1, 2 |
| | Mim | Mim1 | No hMim1 |
| IMS | Small Tims | Tim8-Tim13, no Tim12 Tim9-Tim10 | DDP1, 2, (hTim8), Tim13, Tim9-Tim10 |
| | Mia 40 | Mia40, Erv1 | hMia40, GFER (Erv1) |
| MIM | Tim22 | Tim22, Tim54, Tim9, Tim10 | Tim22, Tim9, Tim10a, b |
| | Tim23 | Tim50, Tim23, Tim44, Tim17 | Tim50, Tim23, Tim44, Tim17a, b |
| | Oxa1 | Mdm38 | Letm1[2] |

[1](Kato and Mihara 2008), [2](Bauerschmitt et al. 2010)

## 1.2.2 Protein transport across and assembly into the outer membrane

The first step in mitochondrial protein import is mediated by a multi-subunit protein-conducting channel located in the outer membrane of mitochondria (Figure 1.1). The "Translocase of the Outer Membrane" TOM acts as the main entry gate for nearly all mitochondrial proteins. It binds mitochondrial preproteins which were synthesized in the cytosol and passes them to the outer membrane protein sorting system SAM or to the inner membrane translocation systems TIM23 and TIM22. The latter two transfer proteins across and into the mitochondrial inner membranes, respectively.

Various receptor proteins within the TOM machinery selectively recognize different substrates. Although some subunits differ or are absent among species all TOM complexes comprise an approximately 40 kDa large protein, termed Tom40, as major component (Macasev et al. 2000; Macasev et al. 2004; Perry et al. 2006; Poynor et al. 2008). The ion conducting property of isolated Tom40 was demonstrated with single channel measurements in lipid membranes (Hill et al. 1998; Ahting et al. 2001; Poynor et al. 2008). Therefore, it is proposed that Tom40 functions as the actual protein-conducting channel in the outer membrane of mitochondria that facilitates the transfer of virtually all mitochondrial pre-proteins synthesized in the cytosol. The components of the TOM complex and their functions are described in detail in chapter 1.3.

The energy source for the transport of proteins across the outer membrane is still a matter of debate. A membrane potential as driving force as it is present across the inner membrane can be excluded due to the constant ion leakage through VDAC in the mitochondrial outer membrane. Furthermore, there is no evidence for ATP hydrolysis coupled to TOM-mediated transport. It is proposed that the transport across TOM and TIM is tightly connected. Presumably, proteins pass passively through TOM until their positively charged target signal is located in the IMS. Then, the preprotein is drawn by the TIM23 complex and inserted into the matrix where components of the import motor in the matrix pull the preprotein through TOM and TIM simultaneously (Endo et al. 2003).

Another model argues that the preproteins could have different binding affinities towards binding sites on both sides of the TOM complex. These so-called trans-binding sites have higher binding affinities towards preproteins than the primary recognition sites of the cytosolic receptors. Preproteins bind to the receptors Tom70, Tom20 and Tom22 on the cytosolic side of TOM and trans-binding sites in the inner-membrane space have been identified on Tom22 and Tom20 (Bolliger et al. 1995; Mayer et al. 1995; Rimmer et al. 2011). The so called "acidic chain hypothesis" describes sequential binding of a targeting signal to strategically situated acidic receptors, the cytosolic domain of Tom20 and the IMS domain of Tom22 which delivers precursors across the outer membrane to Tim23 in the inner membrane (Komiya et al. 1998).

The "Sorting and Assembly Machinery" SAM in the outer membrane is in charge of protein insertion into the outer membrane (Model et al. 2001; Wiedemann et al. 2003). As it is responsible for the "Topogenesis of mitochondrial Outer membrane β-Barrels" it is also called TOB-complex. The components of the SAM complex are Sam50/Tob55, a channel protein, as well as Sam35 and Sam37, two membrane-embedded proteins which are attached to Sam50 (Kozjak et al. 2003). Together with Tom40, Sam50 and Sam35 represent the only proteins essential for cell viability in the outer mitochondrial membrane (Milenkovic et al. 2004; Dolezal et al. 2006). Sam50, forming a β-barrel of most likely 16 β-strands, possesses a so-called polypeptide-transport-associated domain (POTRA-domain) which is a common motif to trigger protein-protein interactions. The SAM complex is in charge for the insertion of β-barrel proteins into the outer membrane after they have been transferred from the TOM complex through the IMS by the small Tim-proteins (Ryan 2004; Gentle et al. 2005). How this insertion of β-barrels into the outer membrane is structurally and energetically mediated is not yet clear. Sam35 binds precursors in a receptor-like manner while Sam37 is responsible for the release of preproteins from the SAM-complex (Chan and Lithgow 2008). Several partner proteins in the IMS assist the SAM complex in the assembly and insertion of preproteins. These partner proteins together with Sam50 form the Mdm-complex for the "Mitochondrial Distribution and Morphology". One protein of this complex, Mdm10, is of importance as it assists in the assembly of the TOM complex.

Eventually, Mim1 (Mitochondrial import 1) is a small protein in the outer membrane which is also taking part in the insertion of proteins into the outer membrane. In close association with the SAM complex, Mim1 was identified to play a fundamental role in the biogenesis of the TOM complex. It has first been discovered in yeast (Mnaimneh et al. 2004) and in *N. crassa* (Schmitt et al. 2006) but homologues in higher eukaryotes have not been found yet. Mim1 has a highly conserved transmembrane region in the C-terminal part, which might be crucial for dimerization while its N-terminus seems to interact with Sam37 (Lueder and Lithgow 2009; Dimmer and Rapaport 2010).

### 1.2.3 Protein transport across and assembly into the inner membrane

The mitochondrial inner membrane comprises two "Translocases of the Inner mitochondrial Membrane" (TIM), the complexes TIM22 and TIM23. It also contains the Oxa complex which was named after its discovery to be responsible for inner membrane insertion of proteins for the "Oxidase Assembly".

The TIM23 complex is responsible for the translocation of all matrix-bound proteins, many inner membrane proteins and also some proteins intrinsically destined to the IMS. The driving force behind the translocation through the inner membrane is the membrane potential ($\Delta\psi$) and the energy carrier ATP. Therefore, two energy sources drive the protein translocation through TIM23: the electrochemical gradient $\Delta\psi$, ensured by the $F_0/F_1$ ATPase and the hydrolysis of ATP in the mitochondrial matrix.

The components of the TIM23 complex can be divided into two groups. First, the membrane-embedded ones, Tim17, Tim21, Tim23 and Tom50, which generate the translocation pore and exploit the energy of the membrane potential for translocation. Second, the matrix-localized proteins, Tim14, Tim16 and Tim44, attached to the complex and the soluble proteins Mge1 and mtHsp70, forming the import motor which pulls precursors by hydrolysing ATP.

The proteins of the TIM23 complex are all highly conserved in the eukaryotic kingdom. All proteins except Tim17 and Tim21 are essential for cell viability in yeast (Bauer et al. 1996). Tim50 exposes a large receptor domain to the IMS which interacts with polypeptide chains coming through the TOM complex. It may also play a role in the regulation of the permeability of the TIM23 complex. Tim21 tightly interacts with the IMS-domain of Tom22 and therefore stabilizes a super-complex intermediate of TOM and TIM23 (Chacinska et al. 2005). One of the channel forming proteins, Tim23, contains a coiled-coil domain of four trans-membrane helices which might be crucial for its dimerization during the import process. It has been shown that the N-terminus of Tim23 can reach into the outer membrane but this association seems to be dynamic and dependent on the translocation activity (Popov-Celeketic et al. 2008). Tim17 acts as a regulator of translocation and is responsible for the sorting of preproteins to the matrix or other complexes of the inner membrane. It consists of four transmembrane helices which are anchored in the inner membrane. The N-terminal part of Tim17 is important for the translocation and may as well play a role in the gating process (Martinez-Caballero et al. 2007). The translocon part of the TIM23 complex, consisting of Tim17 and Tim23, may be capable of inserting protein laterally into the inner membrane without the help of the import motor (van der Laan et al. 2006). Substrate proteins for import motor independent insertion feature a hydrophobic signal sequence followed by a transmembrane helix in the precursor protein.

The import motor complex, consisting of Tim14, Tim16 and Tim44, is attached to Tim23 in the mitochondrial matrix. It drives preproteins into the matrix in an ATP dependent-manner. This process is assisted by two matrix proteins, a nucleotide exchange factor Mge1 and the chaperone mtHsp70. The proteins of the import motor Tim14, Tim16 and Tim44 as well as Mge1 and mtHsp70 are representing the "Presequence translocator Associated import Motor", sometimes called PAM complex. Tim44 is a matrix protein but partially attached to the TIM23 complex in the inner membrane. It comprises two domains, a C-terminal membrane anchor and an N-terminal domain interacting with other proteins from the import motor. It recruits regulatory factors and chaperones and connects the import motor to the translocation channel of the complex. The proteins Tim14 and Tim16 form a stable subcomplex which regulates the activity of mtHsp70, the key player of the import motor (Mokranjac et al. 2006). MtHsp70 pulls polypeptide chains in vectorial transport of ratchet-like binding and release by ATP-hydrolysis. Mutations in mtHsp70 lead to precursors being stuck in the import channel (Scherer et al. 1990). MtHsp70 comprises two domains, an N-terminal ATPase domain and a C-terminal peptide-binding domain which releases the polypeptide chain upon hydrolysis of ATP. The release of ADP is mediated by Mge1, a nucleotide exchange factor.

The TIM22 complex is located in the inner membrane like the TIM23 complex. It consists of a receptor protein Tim54, a channel protein Tim22 and a small protein, Tim18, presumably responsible for the complex assembly. TIM22 is responsible for inner membrane insertion of carrier proteins with six transmembrane helices and other components translocated by the TIM23 complex into the matrix. The driving force for the insertion lies in the membrane potential. The pore forming component Tim22 is homologous to Tim23 (Sirrenberg et al. 1996). The Tim54 protein exhibits a large domain extending into the IMS which might provide a binding site for small Tim-proteins. It is not actively involved in the transport of proteins and plays a more important role in stabilizing the complex (Hwang et al. 2007). Tim18 is a small integral membrane protein of the TIM22 machinery and stimulates the integration of Tim54 into the complex (Wagner et al. 2008).

The Oxa complex was originally found to be responsible for the insertion of subunits of the respiratory chain like Cytochrome b or the $F_0$-sector of the ATP-synthase. Two substrates can be recognized by the complex, *i.e.* nuclear-encoded proteins transported through TIM23 as well as substrates encoded on the mitochondrial DNA. The latter ones are synthesized by mitochondrial ribosomes in the matrix and are translocated co-translationally through binding of mitochondrial ribosomes to the inner membrane protein Oxa1. The recruitment of ribosomes is accomplished by the C-terminal domain of Oxa1 in cooperation with another inner membrane protein, Mba1, which acts as chaperone (Jia et al. 2003; Ott et al. 2006).

### 1.2.4 Protein transport into the inner membrane space

The Mia complex, consisting of Mia40 and Erv1, acts as a disulfide relay system and is involved in release of proteins with cysteine-rich signal sequences into the IMS after they have been imported by the TOM complex. The central component Mia40, also called Tim40, binds these cysteines via disulfide bonds. Mia40 is bound to the inner membrane but has been shown to be functional as a soluble protein in the IMS as well (Naoe et al. 2004). The substrate leaves the Mia complex in an oxidized and almost completely folded state. The sulfhydryl oxidase Erv1, located in the IMS, regenerates Mia40 by reduction of the cysteine side-chains with the help of proteins of the respiratory chain that act as electron acceptors. The Mia complex is also responsible for the formation of the small Tim complexes in cooperation with chaperones. As they assist in the folding of the complexes their return path through TOM is blocked (Herrmann and Kohl 2007).

The transfer of precursor proteins in the inner membrane space is mediated by small TIM-proteins which act as chaperones to guide precursors to their destination (Vial et al. 2002). The small TIM proteins, building the Tim8-Tim13 complex and the Tim9-Tim10 complex, each consist of six low-molecular weight proteins acting together as "guide dogs" for preproteins inserted into the IMS. They connect protein transport from the TOM complex either to SAM or the TIM22 complex. The structures of the small TIMs has been solved to be like a propeller blade (Webb et al. 2006). Both complexes are found in fungus, mammals and plants (see Table 1) while yeast additionally contains Tim12 which is a modified form of Tim10 (Gentle et al. 2007).

### 1.3 TOM Complex: function and components

The first point of contact for all proteins targeted to mitochondria is the TOM complex. In fungi and mammals the TOM complex comprises seven components, which are two primary receptor proteins, Tom70 and Tom20, one secondary receptor Tom22, a protein-conducting channel protein Tom40, and three low molecular-weight proteins, Tom7, Tom6 and Tom5. The single subunits will be described in detail in the following. The TOM core complex of mammals and fungi contains five constituents, Tom40, Tom22, Tom7, Tom6 and Tom5. The TOM complex in plants contains besides the channel protein Tom34 the receptors Tom23/21 and Tom8 and the small proteins Tom7 and Tom6 (Werhahn et al. 2001; Werhahn and Braun 2002; Werhahn et al. 2003; Macasev et al. 2004; Wojtkowska et al. 2005).

The total molecular mass of TOM core and TOM holo complex in detergent solution is described to range between 350 and 500 kDa (Künkele et al. 1998; Werhahn et al. 2003). The TOM complex in plants is smaller with a molecular mass of ~230 kD (Jänsch et al. 1998). Although the subunit composition of the TOM complex among species is remarkably similar, their subunit stoichiometry is still a matter of controversy (Ahting et al. 1999; Schmitt et al. 2005). It is widely accepted that Tom40 forms the channel of the TOM machinery through which precursor proteins thread from the cytosol into the mitochondrial inter membrane space (Hill et al. 1998; Künkele et al. 1998; Ahting et al. 2001). The other subunits are predicted to be attached to Tom40 by single trans-membrane helices. It is not known, however, how many of the other subunits are associated with Tom40.

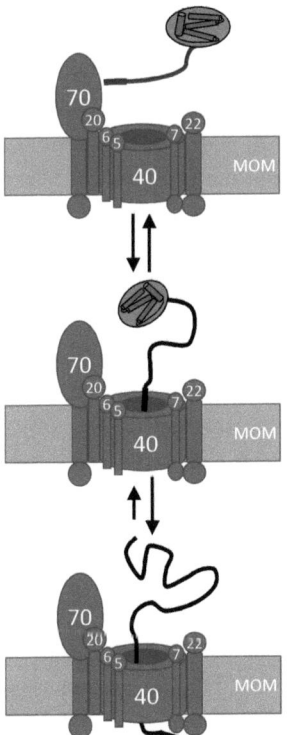

Figure 1.2: Protein transport across outer mitochondrial membrane (MOM) via TOM. The TOM holo complex consists of seven components. The receptor components Tom70, Tom22 and Tom20 are responsible for recognition of mitochondrial precursors; Tom40 represents the main component of the protein-conducting channel. It is believed that mitochondrial preproteins threads through the TOM channel as extended polypeptide chains. The presequence is recognized by the cytosolic domains of the receptors and is passed on to the general import pore. After translocation the presequence is attracted by receptor domains in the intermembrane space.

A first structural view of the multi-subunit core complex was gained from electron microscopy and single particle analysis (Ahting et al. 1999; Ahting et al. 2001; Model et al. 2002; Model et al. 2002; Model et al. 2008). Electron microscopy studies on the TOM core complex have revealed a twin-pore structure with pore diameters of 20 Å (Ahting et al. 1999). This diameter is sufficient to accommodate for unfolded or partially folded mitochondrial preproteins.

Fluorescent correlation spectroscopy studies allowed the analysis of the interactions of mitochondrial presequence peptides with TOM (Stan et al. 2000). Single-channel electrical recordings with purified TOM core complex and Tom40 reconstituted into planar lipid membranes provided first quantitative data of the kinetics of polypeptide interaction (Hill et al. 1998; Ahting et al. 1999; Ahting et al. 2001; Meisinger et al. 2001; Becker et al. 2005; Poynor et al. 2008; Romero-Ruiz et al. 2010). The pore of the TOM complex formed by Tom40 shows cation-selectivity which represents the ideal translocator for positively charged signal sequences of mitochondrial precursors. The TOM complex has presequence binding sites on the IMS (trans) side of the complex, which may involve parts of Tom40 and the IMS domains of Tom20 and Tom7 (Figure 1.2) (Kanamori et al. 1997; Rapaport et al. 1997; Esaki et al. 2004; Yamamoto et al. 2011).

The biogenesis of TOM is mainly dependent on Mim1 located in the outer membrane as it is responsible for the membrane insertion of Tom70 and Tom20 (Becker et al. 2008). Two factors regulate the biogenesis of TOM from the cytosolic side: Casein kinase 2 stimulates the formation of the TOM complex, while protein kinase A inhibits it (Becker et al. 2010; Schmidt et al. 2011). And of course, the precursor Tom proteins require a functional Tom40 pore to enter mitochondria.

### 1.3.1 Tom70

Tom70 is the largest primary receptor in the TOM complex with a molecular weight of approximately 70 kDa in *N. crassa*. It is responsible for the first docking contact with preproteins and then guides them to Tom22 and Tom40 to be translocated through the outer membrane. Tom70 recognizes preproteins with an internal targeting system, which is the case for many multi-transmembrane carrier-proteins of the inner mitochondrial membrane like the ADP/ATP carrier (Söllner et al. 1990; Brix et al. 1997).

The receptor is anchored in the membrane with the N-terminus while the C-terminal cytosolic domain comprises seven tetratricopeptide repeat (TPR) motifs, which are responsible for binding of the chaperones Hsp70 and 90 (Young et al. 2003). The helices in the C-terminal domain are forming a binding pocket for precursor target sequences as well. The structure of Tom70 deduced from crystallographic data is composed of 26 α-helices which mostly show TPR-motifs. The crystal structure of the cytosolic domain indicates that Tom70 is forming a homodimer (Wu and Sha 2006). Due to this dimerization it allows Tom70 to interact with two Hsp70 chaperones simultaneously. Sequence alignments of the amino acid residues responsible for the dimerization show that these residues are conserved between the yeast and the human protein. It remains unclear whether Tom70 is also present as a dimer in the TOM complex. When Tom70 is phosphorylated by cytosolic casein kinases the binding of Hsp70 is suppressed. Therefore, the phosphorylation of receptors of the TOM complex represents a strong tool to regulate transport into mitochondria (Schmidt et al. 2011).

Tom70 has been identified also in early eukaryotes (Tsaousis et al. 2011), however, a Tom70 homologue in plants and algae has not been identified (Chan et al. 2006). In *Arabidopsis thaliana*, a similar protein, mtOM64, seems to replace the receptor function of Tom70 (Chew et al. 2004).

### 1.3.2 Tom20

Tom20 acts as another primary receptor besides Tom70. It is responsible for the binding of precursor proteins with N-terminal signal sequences. It was first identified in yeast where it was shown to act in combination with Tom70 to be responsible for the recognition of subunits precursors of the $F_0/F_1$ ATPase (Söllner et al. 1989). Topological investigations showed that the corresponding receptor Tom23/21 in plants is anchored C-terminally while Tom20 of fungi and mammals is anchored N-terminally in the outer membrane (Perry et al. 2006). The cytosolic domain of Tom20 from both fungi and mammals contains a single TPR motif. Structural analysis of the cytosolic domain by NMR revealed that the C-terminus of mammalian Tom20 forms an α-helical groove to accommodate an α-helix in the presequence structure (Abe et al. 2000). The motif recognized by Tom20 spans only 5-8 amino acids in the target signal. A second binding site at Tom20 has been shown to support the efficiency of import by keeping the precursors close to the complex (Yamamoto et al. 2011). Additionally, Tom20 seems to attract crucial mRNAs for the synthesis of mitochondrial preproteins close to the translocation pore (Eliyahu et al. 2010).

### 1.3.3 Tom22

The secondary preprotein receptor Tom22 (Kiebler et al. 1993) is strongly associated with the general import pore Tom40. This connection was presumably present already in early eukaryotes (Perry et al. 2008). Tom22 receives precursors from the primary receptors Tom20 and Tom70 and guides them to Tom40 to be imported.

Several targeting signals containing a segment of 10–20 residues that fold into a basic amphipathic α-helix have been proven to bind to Tom22 (Rimmer et al. 2011). The receptor is C-terminally anchored in the outer membrane and its N-terminal domain faces the cytosol where it interacts with Tom20 during the protein import (Mayer et al. 1995). Tom22 may promote the dissociation of preproteins from the receptor Tom20 and therefore facilitates the entry of these proteins into the translocation pore. Besides its receptor function, Tom22 plays a fundamental role in the stability of the TOM complex as its deletion results in dissociation into small subcomplexes (van Wilpe et al. 1999). A Tom22 homologue has been identified in human tissue but not in plants (Saeki et al. 2000). In *A. thaliana*, for example, an 8 kDa protein may have similar functions as Tom22 (Macasev et al. 2004).

### 1.3.4 Tom40

The channel protein Tom40 is the only essential component of the TOM complex (Vestweber et al. 1989; Baker et al. 1990; Dekker et al. 1998) and has a mass of around 40 kDa according to the organism. Secondary structure predictions of Tom40s from yeast, fungus, plants and mammals suggest 19 β-strands in the protein structure (Jones 1999). A common motif for all Tom40s is a α-helix located right before the first β-strand. However, in fungus, an additional α-helix is located at the C-terminus behind the last β-strand.

Tom40 receives precursors from the TOM receptors and translocates them through the outer mitochondrial membrane. It is proposed that the inner wall of Tom40 is not entirely hydrophilic but contains some hydrophobic patches (Künkele et al. 1998; Esaki et al. 2003). This presents an optimal environment for the translocation of unfolded polypeptides. The biogenesis of Tom40 into the TOM complex requires mainly the aid of Tom20.

In many species several isoforms of Tom40 have been identified: two human isoforms are known, hTom40A and hTom40B (Humphries et al. 2005), three isoforms in *Bos taurus* (Stutz 2009) and two isoforms in *A. thaliana* (Macasev et al. 2000). Presumably, these isoforms have evolved from gene duplication events. The predicted structures show a highly conserved β-barrel part and a variable elongation at the N-terminus. It is not clear whether different isoforms gather in a hetero-complex or if only one isoform is present in the complex; also whether the isoforms function in a similar way. However, it has been reported that ratTom40B is mainly present in the same tissue as ratTom40A except in testis tissue (Kinoshita et al. 2007).

### 1.3.5 Tom5, Tom6 and Tom7

When the small Tom proteins Tom5 and Tom7 were first discovered, they were entitled according to their molecular weight (Moczko et al. 1992; Hönlinger et al. 1996; Dietmeier et al. 1997). Tom7 of *N. crassa* has a mass of 6.4 kDa while Tom5 is smaller with 5.4 kDa. Tom6 was discovered later (Kassenbrock et al. 1993) and is actually larger than Tom7 with a mass of 7.1 kDa. The function of the small Tom proteins was unclear for a long time as deletion mutants showed only minor defects. It is suggested that they stabilize the complex in *N. crassa* (Sherman et al. 2005) but they seem to play a more stabilizing role in yeast than in *N. crassa* (Schmitt et al. 2005). Tom6 functions as an assembly factor for Tom22, promoting its association with Tom40 and has a stabilizing effect on the complex (Hönlinger et al. 1996; Dekker et al. 1998). Tom7 seems to play a role in recruiting the Mdm10 factor and mediates the assembly of the TOM complex by inserting Tom40 in the complex (Becker et al. 2010). The small proteins Tom5 and Tom6 have not been found in mammals until 2008 when Kato *et al.* showed the existence of the two proteins in human cell tissue. However, in contrast to Tom7 they seem to have a minor effect on the stability of the complex (Sherman et al. 2005; Kato and Mihara 2008)

## 1.4 Beta-barrel membrane proteins

The main component of the TOM complex Tom40 has a predicted β-barrel structure. Many proteins from the outer membrane of Gram-negative bacteria and mitochondria share this common structural motif: a composition of β-sheets forming a β-barrel (Walther et al. 2009). Despite their structural similarity their functions, e.g. uptake of nutrients, diffusion of ion, protein import or enzymatic activity, differ strongly.

The most abundant β-barrel protein in the outer mitochondrial membrane is the voltage-dependent anion channel (VDAC). The protein has a mass of 30 kDa and is not involved in a protein complex although a tendency for dimerization has been reported (Szabo and Zoratti 1993; Keinan et al. 2010). VDAC forms a barrel-like structure of β-sheets which span through the outer mitochondrial membrane. The crystal structure of VDAC1 of two organisms has been solved and reveals a β-barrel protein with 19 β-sheets (Bayrhuber et al. 2008; Hiller et al. 2008; Ujwal et al. 2008) with a resolution of 2.3 Å.

The structure of VDAC with 19 β-strands is remarkable as up to this date no β-barrel protein with an uneven number of β-strands has been identified before. Until the crystal structure of VDAC was solved all structurally known β-barrels were believed to consist of an even number of β-strands. This has been revised with VDAC consisting of 19 β-strands and strand number 1 and 19 connecting in a rather unusual parallel organization.

By comparing the amino acid composition of VDAC and Tom40, there is a similarity of less than 15 %. However, an alignment of the predicted secondary structure of Tom40 with the crystal structure of VDAC shows striking similarity. It seems that both proteins share a similar structure of a β-barrel consisting of 19 β-strands (Figure 1.3). Consequently, it has been hypothesized only recently that VDACs and Tom40s are ancestrally related and should be grouped into the same protein family: the mitochondrial porins (Pusnik et al. 2009; Zeth and Thein 2010). Nevertheless, their functions are diverse, since VDAC is responsible for the ion flux via the outer membrane while Tom40 catalyzes the transport of nuclear-encoded mitochondrial proteins. It is possible that VDAC and Tom40 descent from a common ancestor protein. Therefore, Tom40 is proposed to form a β-barrel, similar to the mitochondrial voltage-dependent anion channel VDAC (Zeth 2010).

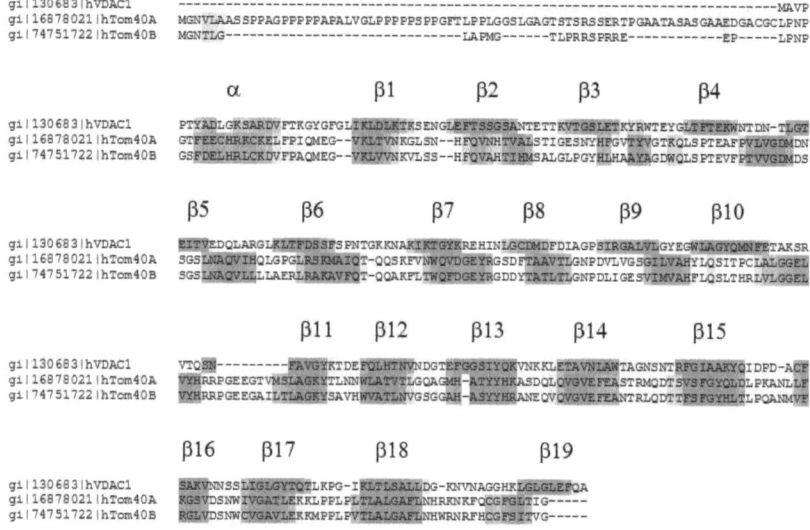

Figure 1.3: Ali2D-predicted secondary structure of full length hTom40 isoform A and B compared with structure from human VDAC1. Ali2D predicted β-sheets are marked in blue, α-helical content is indicated in red, the colour depth is indicating the confidence of the prediction. The regions indicating the β-strands are highly comparable for both human Tom40 isoforms with human VDAC1 suggesting a similar protein structure and evolutionary relation.

This proposed structural relation stands in line with findings about the orientation of Tom40 in the outer membrane. A structural relation to VDAC with an uneven number of β-strands would result in the same orientation of Tom40 N- and C-termini in the membrane. Previous studies already claimed an oppositional orientation of both Tom40 termini (Hill et al. 1998) with the C-terminus facing the intermembrane space (Suzuki et al. 2004) supporting the theory of an uneven number of β-strands and the structural relation to VDAC.

An evolutionary link between pro- and eukaryotic outer membrane proteins has been predicted, but so far has only been confirmed for the integral outer membrane protein Sam50 which is homologous to Omp85, also known as YaeT or BamA (Zeth and Thein 2010). Omp85 belongs to a group of bacterial outer membrane proteins (Omp). These bacterial porins comprise different numbers of β-strands ranging from 8-24. This number is always even-numbered and the C-and N-termini face the periplasm. They have been studied in great detail and the principles of their architecture have been shown first by electron microscopy (Lepault et al. 1988). The first crystal structure of a porin has been solved, describing the porin from *Rhodobacter capsulatus* (Weiss et al. 1990). It has been shown that diffusion porins mainly from trimers and can be divided into two subgroups comprising 16-stranded nonspecific and 18-stranded specific porins (Schirmer et al. 1995; Forst et al. 1998).

Generally spoken, the larger a β-barrel the less stable is the arrangement of β-strands (Das and Matile 2001). Therefore, some β-barrels with an energetically unfavorable conformation exhibit different forms of stabilization. This can either be achieved by the addition of stabilizing structural elements like α-helices or the formation of oligomers. Stabilizing structural elements can be located either inside the barrel where they support the barrel structure from the inside, therefore called "in-plug". Or they can grasp the barrel from the outside and eventually shield weak β-strands, called an "out-clamp" (Naveed et al. 2009).

Mitochondrial β-barrel proteins feature an internal targeting signal termed "β-signal", which is located in the last β-strand. It is not only essential for the protein import across the outer membrane but as well for its correct integration. C-terminally truncated Tom40 is not capable of forming an import intermediate with SAM confirming the position and relevance of a β-signal in the last β-strand of Tom40 (Kutik et al. 2008). Bacterial β-barrel proteins contain a C-terminal signature motif that interacts with the Omp85 complex for membrane insertion. This motif typically consists of 10 amino acids comprising a conserved hydrophobic pattern and a phenylalanine right at the C-terminus (Robert et al. 2006). It is believed that β-barrel proteins did not evolve *de novo* but were built from established structural motifs by duplication and/or recombination of existing protein structures. This method of combining readily folding super-secondary structures like ββ-hairpins to new constructs bears a powerful method for the cell to adapt to a changing environment and is a fundamental process in protein evolution (Söding and Lupas 2003; Arnold et al. 2007).

To date, most Tom40 homologs have been characterized in mitochondria of *Saccharomyces cerevisiae, N. crassa, A. thaliana, Homo sapiens* and *Rattus norvegicus* (Schwartz and Matouschek 1999; Suzuki et al. 2000; Werhahn et al. 2001; Kinoshita et al. 2007). Their structures show a strong structural conservation in the predicted transmembrane domain and a high variability in the extracellular parts. Biophysical and biochemical studies with precursor proteins imported into mitochondria of *S. cerevisiae*, indicated Tom40 pore diameters of ~20 Å (Hill et al. 1998; Schwartz and Matouschek 1999; Suzuki et al. 2000; Ahting et al. 2001; Werhahn et al. 2003; Kinoshita et al. 2007). However, detailed structural and functional studies implying the interaction of preproteins with purified mammalian Tom40 have been hampered by the considerable complexity to purify the protein from native tissue. A promising attempt to study mammalian Tom40 lies in the recombinant expression in bacterial cell culture.

## 1.5 Aim of this study

The aim of this study was to gain further insight into the structure of the TOM complex using biophysical and biochemical methods. The experiments to approach this matter were divided into three parts: (1) Gain insight into the stoichiometry of the TOM core complex and the interaction of its subunits, (2) the high-yield purification of recombinant Tom40 to explore the structural and functional properties of a channel protein and (3) the improvement of stability in a β-barrel to pave the way for studies on the protein translocation channel itself by x-ray crystallography.

(1) Although the subunit composition of the TOM complex of fungi, mammals and plants is remarkably similar, their subunit stoichiometry is still a matter of controversy. Channel characteristics of the TOM complex from *N. crassa* has been studied before in the department of Biophysics so this complex was chosen for detailed analysis on the stoichiometry and subunit interaction. To address this matter I applied a method that had been successfully used to solve the subunit composition of other multi-subunit proteins. Laser induced liquid bead ion desorption coupled with mass spectrometry presents a powerful tool to analyze the subunit composition, stoichiometry and mass of purified TOM core complex.

(2) The structure of a protein provides fundamental information about the function and the interaction mode with other proteins in a complex. In this work the expression of human Tom40 in *Escherichia coli* and the purification of the protein from inclusion bodies under denaturing conditions should be evaluated. Purified human Tom40A and Tom40B should be refolded in detergent solutions for further analysis on their structure and function. Reconstitution into planar lipid bilayers and electrophysiology studies should confirm that both proteins form ion-conducting channels. A base for first 3-D crystallization trials of recombinant human Tom40 should also be established in this work.

(3) To study the interaction of Tom40 with other Tom40 molecules and subunits of the TOM complex, detailed analysis of potential binding sites and stabilizing factors within the protein should be assessed. Weak strands in the Tom40 structure and destabilizing amino acids should be identified and replaced with hydrophobic amino acids by using mutagenesis. Assays concerning temperature sensitivity and folding state in chaotropic reagents should show the increased stability for the mutated protein in comparison with the wild type protein. Furthermore the oligomerization state of the mutated protein should tend to monomers in contrast to the wild type Tom40 to give the base for structural investigation by protein crystallography.

# 2 Materials and methods

## 2.1 Equipment

### 2.1.1 Chemicals

All chemicals have been purchased from Carl Roth (Karlsruhe, Germany), Merck (Darmstadt, Germany) or Sigma-Aldrich (München, Germany) unless otherwise noted. All solutions have been prepared with double-distilled water from a destille (Wagner & Munz, München, Germany) unless otherwise noted.

### 2.1.2 Devices

- Autoclave: Systec 3870 ELV (Systec, Wettenberg, Germany)
- Centrifuges:
  Sorvall Evolution RC, rotors SLA-3000, SA-300, (Sorvall, Langenselbold, Germany)
  Biofuge fresco (Heraeus/Thermo, Langenselbold, Germany)
  Centrifuge 5415D (Eppendorf, Hamburg, Germany)
  Universal 32 (Hettich, Tuttlingen, Germany)
- Ultracentrifuges:
  Beckman, L7-65, rotor Ti70 (Beckman Coulter, Krefeld, Germany)
  Sorvall UltraPro 80, rotor Ti70 (Sorvall, Langenselbold, Germany)
- Magnetic stirrer IKAMAG REO (IKA-Combimag RET, Germany)
- pH-Meter: pH 197i (WTW, Weilheim, Germany)
- Scales CP5202-OCE (Sartorius Göttingen, Germany)
- Precision scale Kern-770 (Kern&Sohn, Balingen, Germany)

All other devices used in experiments for this thesis are specified in the corresponding chapter.

## 2.2 Microbiological methods

### 2.2.1 Bacterial cell culture

Devices:
- Thermoshaker THO 5 (Gerhardt, Bonn, Germany)
- Fermenter BIOFLO 3000 (New Brunswick Scientific, Edison, N.J., USA)

Media:
- $LB_0$-Medium: 10g trypton, 5g yeast extract, 5 g NaCl, $H_2O$ ad 1 L
- Antifoam 405 (Sigma-Aldrich, München, Germany)
- Isopropyl-β-D-thiogalactopyranoside
  (IPTG; Carl Roth, Karlsruhe, Germany)

*E. coli* cultures (see Table 2) were prepared in shaking flasks of variable volume. As a rule of thumb, flasks were filled by 1/5 of their volume with autoclaved LB-medium. Antibiotic stock solutions were added to the medium according to the resistance on the respective plasmid in a dilution of 1:1000. If the culture volume exceeded 1 L, a drop of antifoam was added. To upscale the culture size a fermenter with a volume of 10 L was used. The fermenter and the medium were autoclaved separately and the medium was filled in the fermenter via an autoclaved nozzle. Air supply was regulated to a pressure of 1.4 bar and the temperature was set to either 30 or 37 °C depending on the incubation time. Expression of proteins in pET-vectors was induced with 1 mM IPTG at a cell density corresponding to an $OD_{600}$ of 0.6. Cells were harvested after at least 5 h of growth or, when incubated over night, after 16 h of growth by centrifugation at 2.200 x g at 4 °C for 10 min.

Table 2: *E. coli* Strains

|  | Genotype | Source |
|---|---|---|
| *E. coli* BL21 (DE3) | F<sup>-</sup> *ompT gal dcm lon hsdS*$_B$($r_B^-$ $m_B^-$) λ(DE3) | Stratagene, La Jolla, USA |
| *E. coli* BL21-Codon + (DE3) RIPL | F<sup>-</sup> *ompT hsdS*($r_B^-$ $m_B^-$) *dcm*<sup>+</sup> Tet<sup>r</sup> *gal* λ(DE3) *endA* Hte [*argU proL* Cam<sup>r</sup>] [*argU ileY leuW* Strep/Spec<sup>r</sup>] | Stratagene, La Jolla, USA |
| *E. coli* DH5α | F<sup>-</sup> *endA1 glnV44 thi-1 recA1 relA1 gyrA96 deoR nupG* Φ80d*lacZ*ΔM15 Δ(*lacZYA-argF*)U169, hsdR17($r_K^-$ $m_K^+$), λ– | Invitrogen, Karlsuhe, Germany |
| *E. coli* C41 (DE3) | F<sup>-</sup> *ompT gal dcm hsdS*$_B$($r_B^-$ $m_B^-$)(DE3) | Lucigen Corporation Middleton, WI, USA |
| *E. coli* Top10' | F<sup>-</sup> *mcrA* Δ(*mrr-hsdRMS-mcrBC*) Φ80*lacZ*ΔM15 Δ*lacX74 nupG recA1 araD139* Δ(*ara-leu*)7697 *galE15 galK16 rpsL*(Str<sup>R</sup>) *endA1* λ<sup>-</sup> | Invitrogen, Karlsuhe, Germany |

### 2.2.2 Preparation of chemically competent *E. coli* cells for transformation

Devices:
- Incubator Heraeus (Newport Pagnell, UK)
- Ultra low temperature freezer (-80 °C) U410 Premium (New Brunswick Scientific, Edison, N.J., USA)

Media:
- LB-agar: 10 g trypton, 5 g yeast extract, 5 g NaCl, 7.7 g agar, $H_2O$ ad 1 L

A strain of competent cells was plated on $LB_0$-agar plates and grown over night at 37 °C. For strains used in this preparation refer to Table 2. One clone was picked from this plate and used to inoculate a 10 ml overnight culture. This pre-culture was then used to inoculate a 50 ml culture of $LB_0$-medium. The 50 ml - culture was grown to an $OD_{600} > 0.5$ and then stored on ice for at least 15 min. The cooled culture was centrifuged at 2.200 x g at 4 °C for 10 min in two autoclaved and cooled centrifuge tubes. The supernatants were carefully removed and the pellets were resuspended in 2 x 5 ml ice-cold 100 mM $CaCl_2$ solution. After incubation on ice for 2 h the solution was centrifuged again at 2.200 x g at 4 °C for 10 min. After supernatants have been decanted carefully the pellets were resuspended in 2 x 2.2 ml ice-cold 50 mM $CaCl_2$ with 20 % glycerol. Aliquots of 200 µl were stored in autoclaved eppendorf tubes -80 °C.

### 2.2.3 Transformation

Devices:
- Incubator Heraeus (Newport Pagnell, UK)
- Heating block Thermomixer comfort (Eppendorf, Hamburg, Germany)

Media:
- Kanamycin stock solution: 50 mg/ml in 100 mM NaOH, diluted 1:1000 in $LB_0$-medium or LB-agar
- Ampicillin stock solution: 100 mg/ml in $H_2O$/Ethanol (1:1) diluted 1:1000 in $LB_0$-medium or LB-agar

To transform plasmid DNA into a bacterial strain, chemically competent *E. coli* strains were used. For strains used for transformation refer to Table 2. The transformation process was performed in Eppendorf tubes. The cells were thawed on ice for at least 30 min. Purified plasmid DNA (concentration: ~ 80 ng/µl) was added to the cells and cells were kept on ice for another 20 min. A heat shock of 42 °C was applied in a table top heating block for 1 min, subsequently. Then, 1 ml of $LB_0$-medium was added to the cells and cells were shaken at 37 °C for one hour. Afterwards 100 µl of the cell suspension were plated on agar plates (100 µl spread) containing a selective antibiotic (see 2.2.1). The residual culture was centrifuged and the supernatant was discarded. The cell pellet was solubilized in the remaining drop and plated as well (concentrated spread). The plates were incubated at 37 °C overnight. Normally, the colonies on the 100 µl spread were sufficient for further process. If the transformation process was less efficient, residual colonies from the concentrated spread were used for further experiments.

### 2.2.4 Glycerol stocks

Media:  • Glycerol-stock-medium: 87 % glycerol        57.5 ml
TY-medium (LB$_0$ w/o NaCl) 42.5 ml
autoclave glycerol and TY-medium separately

For long-term storage of *E. coli* cultures, overnight cultures were prepared with 5 ml LB-medium containing the respective antibiotic. The next day, bacterial cells were pelleted by centrifugation at 2.200 x g for 5 min at 4 °C. The supernatant was discarded and the bacterial pellet was solubilized in 1 ml glycerol-stock-medium and stored in autoclaved cap-sealed tubes at -20 °C. Glycerol-stocks could be stored for several years.

### 2.2.5 Isolation of inclusion bodies

Devices:  • French pressure cell, AMINCO (American Instrument Exchange, Haverhill, MA; USA)

Buffers:  • PBS-buffer: phosphate buffered saline, 50 mM Na-phosphate pH 7.2, 100 mM NaCl
• TN-buffer: 50 mM Tris-HCl pH 8, 100 mM NaCl agar
• Guanidin-buffer: 6 M guanidine hydrochloride, 20 mM Tris-HCl pH 8, 150 mM NaCl, 1 mM β-mercaptoethanol
• Lysis-buffer: 50 mM Tris-HCl pH 8, 1 mM EDTA, 100 mM NaCl, 1 mM PMSF, 0.26 mg/ml lysozyme

Inclusion bodies were isolated from *E. coli* cells expressing recombinant Tom40. After cell harvesting by centrifugation (see 2.2.1) cell wall disruption was either performed mechanically with a French press or chemically by cell membrane lysis with deoxycholate.

French Press: *E. coli* cells were washed with PBS-buffer and resuspended in 10 ml PBS-buffer per g cells supplemented with 12.5 U DNaseI from bovine pancreas (Sigma-Aldrich, Hamburg, Germany). After incubation on ice for 20 min cells were lysed in a French press with a pressure of 16.000 – 18.000 psi.

Deoxycholate: *E. coli* cells were washed with PBS-buffer on ice and resuspended in 3 ml lysis-buffer per g cells. 4 mg deoxycholate per g cells was added to lyse membranes and the solution was stirred at 37 °C until it became viscous. After addition of 12.5 U DNaseI followed by an incubation by room temperature for 30 min until the solution became thin fluid.

Both methods for cell wall disruption are followed by a centrifugation at 20.000 x g for 20 min at 4 °C. Afterwards, the inclusion body pellets were washed with TN-buffer and solubilized in guanidine-buffer at a protein concentration of ~ 50 mg/ml using a glass-glass homogenizer. To remove non solubilized material the homogenate was centrifuged at 30.000 x g for 30 min at 4°C and supernatants were recovered and stored at 4 °C for further use.

### 2.2.6 TOM complex isolation from *N. crassa* mitochondria

Devices:
- Corrundum Mill (custom made by scientific workshop, LMU München, Germany)
- Ultracentrifuge: Beckman, L7-65, rotor Ti70 (Beckman Coulter, Krefeld, Germany)

Buffer:
- Solubilization buffer:

| | |
|---|---|
| Glycerol | 20 % |
| Tris-HCl pH 8.5 | 20 mM |
| NaCl | 200 mM |
| PMSF | 1 mM |
| Imidazole pH 8.5 | 20 mM |
| DDM | 1 % |

*N. crassa* cultivation and isolation of mitochondria isolation were performed as previously described (Sebald et al. 1979; Künkele et al. 1998) and improved by Poynor et al. in 2008. Mitochondria were isolated from *N. crassa* strain GR 107 (Künkele et al. 1998) that contains a hexahistidinyl-tagged form of Tom22. Mitochondrial membranes were solubilized for 30 min at 4 °C in solubilisation buffer containing 1 % n-dodecyl-$\beta$-maltoside (DDM) at a protein concentration of 10 mg/ml. The solubilized mitochondria were centrifuged at 100.000 x g at 4 °C for 40 min to pelletize insoluble membrane material. The supernatant was filtered through a 12 µm filter paper and subjected to affinity chromatography.

## 2.3 Molecular biology methods

### 2.3.1 Strains and plasmids

Table 3: Plasmids

| | Source | Remarks |
|---|---|---|
| pET24d hTom40AΔ1-82 | Trenzyme | Kanamycin resistance, C-terminal His-Tag |
| pET24d hTom40BΔ1-29 | Trenzyme | Kanamycin resistance, C-terminal His-Tag |
| pET24d hTom40AΔ1-82 K27L | this work | Kanamycin resistance, C-terminal His-Tag |
| pET24d hTom40A Δ1-82 H37L | this work | Kanamycin resistance, C-terminal His-Tag |
| pET24d hTom40A Δ1-82 H140L | this work | Kanamycin resistance, C-terminal His-Tag |
| pET24d hTom40A Δ1-82 K27L, H37L | this work | Kanamycin resistance, C-terminal His-Tag |
| pET24d hTom40AΔ1-82 K27L, H37L, H140L | this work | Kanamycin resistance, C-terminal His-Tag |
| pET24d btTom40A | this work | Kanamycin resistance, C-terminal His-Tag |
| pET24a atTom40-1 | this work | Ampicillin resistance, |

Table 4: Primers

| | Sequence 5'→3' |
|---|---|
| for hTom40A K27L | GGGTGTCAAGCTCACAGTCAATCTAGGGTTGAGTAACCATTTTCAGGTCAAC |
| rev hTom40A K27L | CCTGAAAATGGTTACTCAACCCTAGATTGACTGTGAGCTTGACACCCTCCATC |
| for hTom40A H37L | GAGTAACCATTTTCAGGTCAACCTCACAGTAGCCCTCAGCACAATCGGG |
| rev hTom40A H37L | GTGCTGAGGGCTACTGTGAGGTTGACCTGAAAATGGTTACTCAACCC |
| for hTom40A K27L, H37L | TCAATCTAGGGTTGAGTAACCATTTTCAGGTCAACCTCACAGTAGCCCTCAGCACAATCGG |
| rev hTom40A K27L, H37L | CTGTGAGGTTGACCTGAAAATGGTTACTCAACCCTAGATTGACTGTGAGCTTGACACCCTCCATC |
| for hTom40A H140L | GGGTTCAGGAATCCTCGTAGCCCTCTACCTCCAGAGCATCACGCCTTGCCTGGC |
| rev hTom40A H140L | GGCAAGGCGTGATGCTCTGGAGGTAGAGGGCTACGAGGATTCCTGAACCCACGAGG |
| for btTom40A | ATATATCCATGGGGAACGTATTGGCCGCTAGCTCGC |
| rev btTom40A | TAATTACTCGAGACCAATGGTGAGGCCGAAGCCACACTGGAAC |
| for atTom40-2 | ATATATCATATGGAGGGCTTTTCACCACCGATTAACACTGCG |
| rev at Tom40-2 | ATATAACTCGAGAAAGGCGTTAACACCGAAACCAAACTTGTAATCC |
| T7 prom | TAATACGACTCACTATAGGG |
| T7 term | TATGCTAGTTATTGCTCAG |
| SP6 | ATTTAGGTGACACTATAG |
| pET-RP | CTAGTTATTGCTCAGCGG |

All primers were purchased in HPLC grade from Sigma-Aldrich except the four primers T7 prom, T7 term, SP6 and pET-RP which are commercially available and were only used for sequencing.

## 2.3.2 Isolation of plasmid DNA

Devices:
- Nanodrop UV Spectrometer
  (Thermo Fischer Scientific, Wilmington. DE, USA)
- peqGOLD Plasmid Miniprep Kit I (Safety Line)
  (Peqlab, Erlangen, Germany)

Isolation of plasmid DNA from *E. coli* cells has been done according to a standard protocol included in the peqGOLD Plasmid Miniprep Kit. The favored bacterial strain was plated on selective agar plates and incubated over night at 37 °C. Bacterial cell cultures were inoculated with one clone from an agar plate in a volume of 5 ml and were shaken at 37 °C overnight. The bacterial cells were pelleted by centrifugation at 4.000 x g at 4 °C for 5 min. The pellet was resuspended in buffer from the kit and the isolation procedure was performed as advised from the manual. Eventually, plasmid DNA was eluted from the spin columns with autoclaved $ddH_2O$. The DNA concentration of the plasmid solution has been determined by UV absorbance spectroscopy at 260/230 nm using a Nanodrop UV Spectrometer. Plasmid DNA was stored at -20 °C.

## 2.3.3 Agarose gel electrophoresis

Devices:
- Gel chamber B2 (Peqlab, Erlangen, Germany)
- Geldocumentation system: Chemidoc, XRS (Biorad, München, Germany)
- Transilluminator TFX-35M with digital camera FAC 831
  (Bioblock Scientific, Illkirch, France)

Buffer:
- 50 x TAE running-buffer: Tris-HCl        242 g
  Glacial acetic acid        57.1 ml
  EDTA 0.5 M pH 8        100 ml

Qualitative analysis of plasmid DNA was determined with agarose gel electrophoresis. The gels contained an agarose concentration between 0.8 and 1.2 % according to the size of the fragments to be analyzed. Ethidiumbromide was added directly to the gel in a concentration of 50 µM (2 µl per 100 ml agarose solution). If the gel was only for analytical purpose the DNA sample had a volume of 3-5 µl. If the DNA-sample should de recovered for further ligation at least 10 µl have been loaded in the gel pocket. Samples were mixed with loading buffer (6 x DNA Loading Dye, Fermentas, St. Leon-Rot, Germany). The gel was placed in a running chamber containing 1 x TAE-buffer. The running conditions were 5 V $cm^{-1}$ electrode distance, so 120 mV for large and 80 mV for small gel chambers was applied, respectively. Afterwards, gels have been visualized on an UV-table and a gel picture was taken for documentation.

### 2.3.4 DNA extraction from agarose gels

Device:  • peqGOLD gel extraction kit (Peqlab, Erlangen, Germany)

To recover plasmid DNA from preparative agarose gels bands were visualized shortly on an UV table to minimize UV damage on DNA strands. The desired bands were cut out with a clean scalpel and transferred to an Eppendorf tube. DNA extraction from agarose pieces was performed following the procedures described in the manual of the peqGOLD gel extraction kit. DNA was eluted from spin columns with autoclaved ddH$_2$O and stored at -20 °C.

### 2.3.5 DNA Sequencing

DNA-sequencing was performed by AGOWA (Berlin, Germany), GATC (Konstanz, Germany) or by the in-house sequencing service at the Max-Planck-Institute for Developmental Biology (Tübingen, Germany). All sequences, supplied online, were analyzed with the freeware software package "Chromas Lite" (Technelysium Pty Ltd, 2005).

### 2.3.6 Digestion of plasmid DNA

Plasmid DNA was digested with digestion enzymes from New England Biolabs (NEB) or Roche Diagnostics. In a reaction volume of 10 µl, 1 µl of plasmid DNA (~ 80 ng/µl) was added as well as buffer and BSA according to instructions of the companies NEB or Roche. The restriction enzyme was added last at a concentration of ~10-50 U. Reactions were incubated at 37 °C for at least 1 h and digested plasmids were analyzed by agarose gel electrophoresis (see 2.3.3).

### 2.3.7 Ligation

To ligate digested DNA fragments into linearized expression vectors the enzymatic properties of the T4-DNA ligase (Promega, Mannheim, Germany) were utilized in a concentration of ~5 U. In a double reaction ~ 80 ng vector plasmid DNA was added to insert plasmid DNA at a concentration of ~40 ng and ~240 ng, respectively. The reaction volume was filled up to 10 µl with 1 µl of ligation buffer and autoclaved ddH$_2$O. The reaction mix was incubated for 50 min at room temperature and then heated to 37 °C for 10 min. Ligated plasmids were directly used for transformation in *E. coli* Top10' strains.

## 2.3.8 PCR and site-directed mutagenesis

Device: • PCR Mastercycler personal, (Eppendorf, Hamburg, Germany)

Software: • Clone Manager Suite 7 (Sci-Ed Software, Cary, NC USA)

The PCR reactions were prepared in a volume of 50 µl. The primers (Table 4) were added in a volume of 1 µl and plasmid DNA (~ 80 ng/µl) was diluted 1:50, when it was taken from plasmid isolation. The final plasmid concentration in the reaction mix was 2 ng. The DNA polymerases Pfu Ultra II Hot Start (Agilent Technologies, Inc. Santa Clara, CA USA) or Phusion High Fidelity (NEB, Frankfurt, Germany) were added in a concentration of 10 U with the respective buffer. The melting temperature of the primers was calculated with the software Clone Manger. The annealing temperature was calculated from the average melting temperature subtracted by 5 °C. Denaturation, annealing and primer elongation was repeated in 20 cycles.

| PCR-reaction | | | |
|---|---|---|---|
| | ddH$_2$O | ad 50 µl | |
| | 5x / 10x buffer | 10 / 5 µl | |
| | dNTP-mix | 1 µl | 40 µM |
| | forward primer | 1 µl | |
| | reverse primer | 1 µl | |
| | plasmid DNA | 1 µl | ~2 ng |
| | Polymerase | 0.5 µl | ~ 10 U |
| **PCR-program** | Initial denaturation | 95 °C | 30'' |
| | Denaturation | 95 °C | 30'' ⎤ |
| | Annealing | 55 °C | 1'  ⎬ x 20 |
| | Primer elongation | 68 °C | 8'  ⎦ |
| | Final polymerization | 72 °C | 1' |

To replace specific amino acids in a coding region of plasmid DNA, primers with the nucleotide mutation were designed with the software Clone Manager. The primer length was restricted by an upper annealing temperature of 68 °C and the tendency to form hairpins has been minimized by rearranging guanins or cytosins at both ends of the primer. The polymerases Pfu Ultra II and Phusion were used for the PCR reaction. Primers for the site-specific mutagenesis in hTom40AΔ1-82 are listed in Table 4. After the PCR reaction the samples were digested with the restriction enzyme DpnI, which digests the parental plasmid without the mutation only. 1 µl of DpnI (20 U, NEB) was added right to the reaction mix and incubated for 3 h at 37 °C.

## 2.4 Biochemical Methods

### 2.4.1 Determination of protein concentration

The concentrations of protein samples were determined either with the Bradford method (Bradford 1976) for isolated mitochondria or TOM complex. The protein concentration of all recombinantly expressed proteins was assessed with a nanodrop UV spectrometer.

#### 2.4.1.1 Bradford

Device: • Photometer Ultrospec II, LKB Phamacia (Freiburg, Germany)

For determination of the protein concentration in solution according to Bradford a standard curve was prepared for each measurement by diluting a stock solution of bovine gamma globuline (2 mg/ml) in different concentrations to the same buffer as the protein to be analyzed. According to the approximated concentration the protein was diluted in a final volume of 24 µl. Bradford Solution (BioRad Proteinassay, BioRad, München, Germany) was diluted 1:5 with $ddH_2O$ and 1 ml was added to the standard reagents and the protein samples. After 10 min incubation at room temperature the absorbance at 595 nm of the standard and the protein samples were measured in a photometer. The protein concentration in the samples was determined according to the standard curve. This method was used for determination of protein concentration of the TOM complex fractions.

#### 2.4.1.2 Nanodrop

Devices: • Nanodrop UV Spectrometer
(Thermo Fischer Scientific, Wilmington, DE, USA)

The nanodrop UV Spectrometer is capable of determining a protein concentration in a volume of 1-2 µl without processing the solution. After equilibration with the respective buffer the protein concentration can be measured at a wavelength of 280 nm taking the molecular mass and the extinction coefficient into account. Both values were calculated with Protparam from the Swiss Institute of Bioinformatics (Gasteiger 2005).

### 2.4.2 SDS-PAGE

Devices: • Gel pouring bracket (BioRad, München, Germany)
• Running chamber: Mini Protean III (BioRad, München, Germany)

The denaturating sodiumdodecylsulfate polyacrylamide gel-electrophoresis (SDS-PAGE) is a separation of proteins by mass and is based on a method developed by Lämmli (Laemmli 1970).

### 2.4.2.1 SDS-PAGE gel preparation

Acrylamide gels with a size of 6 x 8 cm and a thickness of 0.75 mm were prepared using Mini Protean III brackets from BioRad. Combs for 10 - 15 loading pockets were used. All devices were cleaned with 70 % ethanol prior to use. Stacking gel (4 %) and separating gel (14 %) were prepared as given in Table 5.

Table 5: SDS-PAGE gel preparation

|  | Separating Gel | Stacking Gel |
|---|---|---|
| Acrylamide/Bisacrylamide 30 %/0.8 % | 4.7 ml | 0.7 ml |
| Gelbuffer* | 2.5 ml | 1.25 ml |
| 10 % (w/v) SDS | 100 µl | 50 µl |
| $H_2O$ | 2.7 ml | 3 ml |
| TEMED | 100 µl | 50 µl |
| 10 % (w/v) APS | 10 µl | 10 µl |

* 0.5 M Tris-HCl pH 6.8 for the stacking gel, 1.5 M Tris-HCl pH 8.8 for the separating gel

After pipetting the solutions for the separating gel (Table 5) the mixture was poured between the glass plates and coated with isopropanol to prevent oxidation and remove air bubbles. When the polymerization was finished the isopropanol was removed and the stacking gel (Table 5) was poured on top. The gels have been packed in wet tissues and were stored for up to two weeks.

### 2.4.2.2 Sample Preparation

Buffers:
- 4 x Lämmli-buffer:

| | | |
|---|---|---|
| | Glycerol | 40 % (v/v) |
| | Bromphenolblue | 0.04 % (w/v) |
| | β-Mercaptoethanol | 4 % (v/v) |
| | Sodiumdodecylsulfate | 8 % (w/v) |
| | Tris-HCl pH 6.8 | 0.25 M |
| | (Aliquots stored at -20 °C) | |

- 1 x Lämmli-buffer:

| | | |
|---|---|---|
| | 4 x Lämmli-buffer | 1 Vol. |
| | $ddH_2O$ | 2 Vol. |
| | Tris-HCl pH 6.8, 0.5 M | 1 Vol. |
| | (stored at 4 °C) | |

Protein samples for SDS-PAGE were mixed with 4 x Lämmli-buffer and boiled in a heat block at 96 °C for 5 min. To recover condensed water from the lid, the samples were spun down in a benchtop centrifuge. If protein samples contained guanidine-hydrochloride, an ethanol precipitation was done prior to SDS-PAGE by mixing 100 µl protein solution with 400 µl of ice-cold ethanol (-80 °C). The solution was vortexed and spun down at 16.000 x g for 20 min. The supernatant was removed completely with a pipet and the protein pellets were dried completely in an exsikkator. The pellets were then solubilized in 1 x Lämmli-buffer and proteins were denatured at 96 °C for 5 min.

### 2.4.2.3 Running conditions

| Buffer: | • 10 x SDS running-buffer: | Tris-HCl | 248 mM |
|---|---|---|---|
| | | Glycine | 1.92 M |
| | | Sodiumdodecylsulfate | 1 % (w/v) |

A protein marker with standard proteins has been added to the first lane of the gel. A voltage of 200 mV and a current of 25 mA were applied. The running time was about 1.2 h. When the running front had reached the end of the glass plates the gel was removed and either stained or blotted.

### 2.4.2.4 Tricine-PAGE

| Devices: | • Large running chamber (custom made by scientific workshop LMU München, Germany) |
| | • Running chamber: Mini Protean III (BioRad, München, Germany) |

| Buffers: | • AB-Mix (Hunte 2003): | Acrylamide | 48 g |
|---|---|---|---|
| | | Bisacrylamide: | 1.5 g |
| | | ad H$_2$O | 100 ml |
| | • 3 x Gelbuffer: | Tris-HCl pH 8.45 | 3 M |
| | | SDS | 0.3 % |
| | • 10 x Anode buffer: | Tris-HCl pH 8.45 | 1 M |
| | • 10 x Cathode buffer: | Tris-HCl pH 8.25 | 1 M |
| | | Tricine | 1 M |
| | | SDS | 1 % |

For a separation of small proteins in the range of 1 – 30 kDa Tricine SDS-PAGE was used for protein analysis (Schägger and von Jagow 1987). The gels were poured as noted above or, for a larger separation range, poured in custom-made casks with a size of 16 cm x 14 cm and a gel thickness of 1 mm. For the large gels combs for 14 loading pockets were used. For large gels a bottom gel was poured as well to ensure tight sealing for gel preparation.

Table 6: Tricine-PAGE gel preparation

|  | Bottom Gel | Separating Gel | Spacer Gel | Stacking Gel |
|---|---|---|---|---|
| AB-Mix | 8.7 ml | 10 ml | 6 ml | 1 ml |
| 3 x Gelbuffer | 3.7 ml | 10 ml | 10 ml | 3 ml |
| 40 % Glycerol | - | 7.5 ml | 7.5 ml | - |
| $H_2O$ | - | 30 ml | 30 ml | 12 ml |
| TEMED | 25 µl | 100 µl | 150 µl | 90 µl |
| 10 % (w/v) APS | 50 µl | 10 µl | 15 µl | 9 µl |

The maximum sample volume for Tricine SDS-PAGE was 30 µl and sample preparation was done as described in 2.4.2.2. The running conditions for small tricine gels were 30 mV at 30 mA until the samples completely entered the stacking gel. Then, voltage and current were increased to 200 V at 100 mA which resulted in ~ 3 h running time. Visualization of gel bands with coomassie or silver staining has been performed as described in 2.4.2.5 and 2.4.2.6. Western Blotting required a blotting time of 1 h at 20 V and 200 mA.

### 2.4.2.5 Coomassie staining of SDS-PAGE gels

Solutions:
- Coomassie-Staining solution:
  - Coomassie $R_{250}$    0.2 % (w/v)
  - Coomassie $G_{251}$    0.05 % (w/v)
  - Ethanol    42.5 % (v/v)
  - Methanol    5 % (v/v)
  - Acetic acid    10 % (v/v)
- Destaining solution:
  - Ethanol    26 % (v/v)
  - Acetic acid    8 % (v/v)
- Preserving solution:
  - Acetic acid    7.5 %

The gels were stained for ~30 min with Coomassie staining solution and destained until the protein bands were clearly visible. To preserve the gel and destain the background completely gels were kept in preserving solution overnight. For documentation the gels were scanned in a flatbed scanner and eventually dried using a vacuum drier at 65 °C.

### 2.4.2.6 Silver Staining of SDS-PAGE gels

Devices:
- Silver Stain Kit (Fluka, München, Germany)
- Shaker GFL 3005 (DJB Labcare, Buckinghamshire, UK)

Silver staining of SDS-gels was performed according to a standard protocol of the Silver Stain Kit for proteins. This method has been used when a very sensitive protein band staining of SDS-gels was necessary.

### 2.4.3 Immunoblotting of proteins

| Buffers: | | | |
|---|---|---|---|
| • Transfer buffer: | 1 x SDS-running buffer | 400 ml | |
| | 10 % Methanol | 100 ml | |
| • 10 x TBS: | Tris-HCl pH 7.6 | 500 mM | |
| | NaCl | 1.5 M | |
| • Blocking buffer: | Milk powder | 5 % (w/v) | |
| | in TBS | 1 x | |
| • Ponceau solution: | Ponceau Red | 0.5 % (w/v) | |
| | acetic acid | 1 % (v/v) | |
| • TBS-Tween: | TBS | 1 x | |
| | Tween 20 | 0.1 % | |
| • AP-Buffer: | Tris-HCl pH 9.6 | 100 mM | |
| | NaCl | 100 mM | |
| | $MgCl_2 \times 6\ H_2O$ | 5 mM | |

Devices:
- Blotting apparatus (custom made by scientific workshop LMU München, Germany)
- Developer P1000 (Agfa, Mortsel, Belgium)

To detect residual amounts of protein in SDS-gels the gels were blotted on a membrane. In this case a prestained protein marker has been used for the SDS-PAGE. Depending on the method of detection either a cellulose- or PVDF-membrane was used. The blotting apparatus was custom-made. The gels were placed on the cathode plate with 3 whatman-papers below which have been soaked in transfer buffer and coated again with three soaked whatman-papers. Air bubbles were removed by rolling a glass pipette over the layers. The lid with the anode plate was placed on top and a voltage of 20 mV with a current of 100 mA was applied. To increase the blotting efficiency, several lead cubes were placed on top of the apparatus. The running time was about 50 min. Afterwards the gel was stained with Coomassie staining solution to check the efficiency of protein transfer. The membrane was stained with Ponceau Red to detect protein bands in case parts of the membrane should be decorated with different antibodies.

The membrane was decorated with polyclonal primary antibodies (Table 7) which were diluted 1:1000 from sera in blocking buffer for 1 h. The first antibody was recovered afterwards and kept at 4 °C for further use. To remove nonspecific bound antibodies the membrane was rinsed with TBS-Tween for 6 x 5 min. The second antibody binds specific to the first antibody was used in a dilution of 1:10.000. The membrane was incubated for 1-2 h and then washed with TBS-tween for 3 x 5 min and TBS for another 3 x 5 min to remove nonspecific bound antibodies. Tween20 had to be removed to ensure binding between the substrate and the alkaline phosphatase. Prior to substrate reaction the membrane was rinsed with water.

The detection of antibodies has been performed with CDP-star (Roche, Mannheim, Germany) diluted 1:100 in AP-buffer. 500 µl of this solution were incubated on the membrane in plastic foil for 5 min. The chemiluminescence reaction was visualized on x-ray photo films (Super RX, Fujifilm, Düsseldorf, Germany) in a time range of 30 sec to 10 min with a developer according to supplier instructions.

Table 7: Antibodies

|  | Source | Remarks |
| --- | --- | --- |
| Rabbit-α-Tom40 | Santa Cruz Biotechnology, Heidelberg | For mammalian Tom40 |
| Rabbit α-NcTom40 | LMU München | - |
| Rabbit α-NcTom 22 | LMU München | - |
| Rabbit α-NcTom 20 | LMU München | - |
| Rabbit α-NcTom 70 | LMU München | - |
| Rabbit α-NcTom 6 | LMU München | Blocking with BSA |
| Rabbit α-NcTom 5 | LMU München | - |
| Rabbit α-NcTom 7 | LMU München | - |
| Goat-α-VDAC1 | Santa Cruz Biotechnology, Heidelberg | - |

### 2.4.4 Cross-Linking

For cross-linking experiments 40 µg of protein in a volume of 95 µl 20 mM Na-Phosphate buffer pH 8 was incubated with 5 µl freshly prepared glutaraldehyde (Roth, Karlsruhe, Germany) with a final concentration of 125 µM at 37 °C for 0-

45 min. Cross-linking reactions were stopped by the addition of 10 µl Tris-HCl pH 8 in a final concentration of 50 mM. Aliquots were removed before and after the addition of the cross-linking reagent after certain reaction times. Crosslinking educts were directly analyzed by SDS-PAGE and Western blotting.

### 2.4.5 Protein chromatography

Device:     • ÄKTA basic P900 (GE Healthcare, Uppsala, Sweden)

Software:   • Unicorn 4.12 (GE Healthcare, Uppsala, Sweden)

Columns:    • Ni-Sepharose HisTrap HP 1-20 ml (GE Healthcare, Uppsala, Sweden)
            • ResourceQ 1 and 6 ml (GE Healthcare, Uppsala, Sweden)
            • Superose 6 and 12 (GE Healthcare, Uppsala, Sweden)
            • Superdex 75 and 200 (GE Healthcare, Uppsala, Sweden)

For protein chromatography all buffers used on the ÄKTA chromatography system were filtered through nitrocellulose filters with a pore size of 0.2 µm (Sartorius Stedim Biotech GmbH, Göttingen, Germany) with a vacuum pump or centrifuged at 16.000 x g for 10 min of the sample volume was below 1 ml. All purifications were carried out at 4 °C. The purity of purified proteins was assessed by SDS-PAGE followed by Coomassie or silver staining (see 2.4.2). Protein concentrations were determined using the method of Bradford or the UV spectroscopy using a nanodrop UV spectrometer (see 2.4.1). For storage all columns were washed with 2 CV of water and 2 CV of 20 % ethanol and stored tightly closed at 4 °C.

#### 2.4.5.1 Affinity chromatography for TOM core complex and Tom40 from *N. crassa*

TOM complex was solubilized from *N. crassa* mitochondrial membranes (see 2.2.6) and subsequently purified with Ni-NTA-chromatography via the hexahistidinyl-tag at Tom22. The solubilized outer membrane proteins were passed through a Ni-Sepharose HisTrap equilibrated with 20 mM Tris-HCl pH 8.5, 1 mM PMSF, 10 % glycerol, 300 mM NaCl and 0.1 % DDM. After loading, the resin was washed with buffer without salt containing 20 mM imidazole to remove nonspecifically bound proteins. If Tom40 was to be eluted buffer containing 1.5 % n-octyl-β-glucopyranoside (β-OG), instead of DDM, was used to separate Tom40 from the TOM complex. Otherwise the whole complex was eluted with buffer containing 300 mM imidazole. The column was washed with buffer containing 1 M Imidazole for cleaning and removing other bound proteins.

### 2.4.5.2 Affinity chromatography for recombinant Tom40

Recombinant Tom40 was passed onto Ni-Sepharose HisTrap columns of variable bed volume (1-20 ml) according to the amount of protein refolded. The purification of recombinant Tom40 was either under denaturing conditions with buffer containing 6 M guanidine-hydrochloride or under refolded conditions with buffer containing detergent. The detergent of choice was 0.1 % LDAO (see 2.4.7) or varied upon subsequent experimental requirements. Additionally, the buffers used for the purification usually contained 20 mM Tris-HCl pH 8, 1 mM β-mercaptoethanol (β-ME), 150 mM NaCl for buffer A. Buffer B also contained 1 M imidazole.

After the protein solution had been loaded, the column was washed with a least 2 column volumes (CV) of buffer A until the absorption reached a constant level. Nonspecifically bound proteins have been removed with 20 mM imidazole in a step gradient. The elution of the hexahistidinyl-tag bound protein hTom40 was effective with 300 mM imidazole. Eventually the column was washed with 1 M imidazole.

### 2.4.5.3 Size exclusion chromatography for TOM core complex and Tom40

Further purification of proteins after affinity chromatography was achieved by size exclusion chromatography (SEC). The columns used were all produced by GE Healthcare and included Superose 6 and 12 as well as Superdex 75 and 200 according to the size of the protein.

The columns were equilibrated with 1.5 CV of buffer and protein samples were applied and eluted with the same buffer. The buffer for SEC of the TOM complex contained 20 mM Tris-HCl pH 8.5, 1 mM PMSF, and 0.03 % DDM. SEC buffer for recombinant Tom40 contained 20 mM Tris-HCl pH 8, 1 mM β-ME, 150 mM NaCl and detergent. Protein samples with a volume of 500 µl were centrifuged before application to the column to remove possible aggregates from the protein solution. If necessary, fractions of the eluate were pooled and concentrated using spin columns with an appropriate cut-off (Vivaspin, GE Healthcare).

### 2.4.5.4 Ion Exchange chromatography for TOM core complex from *N. crassa*

For further purification of the TOM complex after affinity chromatography (see 2.4.5.1) TOM complex-containing fractions were pooled and transferred to a Resource Q anion exchange column equilibrated with buffer A containing 20 mM Tris, pH 8.5, 2 % DMSO and 0.1 % DDM and buffer B containing 1 M KCl, additionally. TOM core complex was eluted with a step gradient of 0 – 1 M KCl. Weakly bound proteins were eluted at 200 mM KCl, TOM core complex eluted at 400 mM KCl and tightly bound proteins were eluted at 1 M KCl.

## 2.4.6 Stripping and recharging of Ni-Sepharose HisTrap columns

Ni-Sepharose HisTrap columns were recovered after 4-6 purifications according to standard protocols from GE Healthcare. Ni-ions were stripped off the resin with 20 mM sodium-phosphate, 500 mM NaCl and 50 mM EDTA pH 7.4. The column was washed with 20 % isopropanol and recharged with 100 mM $NiSO_4$. The $NiSO_4$-flowthrough and the subsequent wash with water were collected separately. This recharging process resulted in perfect purifying performance and, if no clogging of occurred, columns could be recharged for several times.

## 2.4.7 Refolding screen

Membrane proteins recombinantly expressed in *E. coli* are often deposited in inclusion bodies where they form aggregates. To restore these proteins to their native state they were purified under denaturating conditions in guanidine-hydrochloride (see 2.4.5.2) and eventually refolded by dilution of purified protein into detergent solution.

To test the efficiency of refolding regarding choice of detergents and pH a refolding screen was performed including four different pH and six different detergents. The buffer strength was set to 50 mM and the detergent concentration was set to 5 x critical micellar concentration (CMC). To guarantee a reduced state 1 mM β-ME was added. Human Tom40 in 6 M guanidine hydrochloride with a concentration of ~6 mg/ml the protein was diluted 1:20 in different refolding buffer solutions at 4 °C. The buffer substances used in this screen were citric acid at pH 5, sodium-phosphate-buffer at pH 6, Tris-HCl-buffer at pH 7 and glycine-sodium-hydroxide at pH 10. The detergents were either DDM, lauryldimethylamine-oxide (LDAO), n-octyl-polyoxyethylene (oPOE), Brij35, 3-[(3-cholamidopropyl-)dimethylammonio]-1-propanesulfonate (CHAPS) and β-OG. The solutions containing refolded protein were vortexed and kept at 4 °C. After one hour the efficiency of protein folding was assessed by centrifugation at 15.000 x g for 10 min at 4 °C and evaluation of the amount of precipitated protein. Additionally, the protein concentration of the supernatant was determined at 280 nm. To check the stability of the protein in detergent solution the concentration determination was repeated after one day and one week.

Optimal refolding was achieved by a tenfold dilution of denatured Tom40 (5 mg/ml in buffer containing 6 M guanidine hydrochloride) into 20 mM Tris-HCl pH 8, 150 mM NaCl, 1 mM β-ME and 0.5 % (w/v) LDAO at 4 °C. To verify the efficient refolding process final samples were centrifuged at 100.000 x g and supernatants were subjected to affinity chromatography to purify and concentrate the refolded protein (see 2.4.5.2).

### 2.4.8 Concentration and dialysis of protein samples

The concentration of purified proteins was determined as described in 2.4.1 and if necessary the protein solution was concentrated. This was done with concentrator tubes with a cut-off from 5-30 kDa in respect of protein mass. As not only the protein solution but also buffer components like detergents and glycerol are concentrated all samples were dialyzed for the following experiments. The dialysis tubes had a cut-off of 8 or 25 kDa and were boiled in 10 mM EDTA for 10 min prior to use. The dialysis tubes were stored in 10 mM EDTA and 50 % ethanol at 4 °C.

## 2.5 Biophysical and structural methods

### 2.5.1 Dynamic light scattering

Devices: • Dynamic Light Scatterer Zetasizer Nano-ZS
(Malvern Instruments, Worcestershire, UK)

Software • Zetasizer Software 6.01 (Malvern Instruments, Worcestershire, UK)

Dynamic Light scattering (DLS) was used to determine the hydrodynamic radius, polydispersity and the presence of aggregates in protein samples containing recombinant Tom40. Protein samples were prepared as described in 2.4.5.2 and a protein concentration of ~0.2 mg/ml was sufficient for analysis. A volume of 15 µl was needed to fill the cuvette and could be recovered afterwards. The measurements were performed at a wavelength of λ = 633 nm at 20 °C. The laser intensity of a 4 mW He-Ne laser was automatically adjusted to the samples properties.

From the correlation function the diffusion coefficient (D) of the molecules was calculated by fitting the data. Finally the hydrodynamic radius ($R_h$) of the particles and molecules was calculated with the software Zetasizer Software 6.01 provided by Malvern Instruments with the function $D = kT/(6\pi\eta_0 R_h)$ where k is the Boltzmann-constant, T is temperature and $\eta_0$ is the solvent viscosity (Viscotek Europe Ltd., Malvern Instruments, Worcestershire, UK). Calculated hydrodynamic radii represented an approach to the actual size of the particles.

### 2.5.2 CD spectroscopy

Device: • CD spectrometer Jasco J.815 (Jasco Inc., Tokyo, Japan)

software: • Spectra Manager Version 2.06.00 (Jasco Inc., Tokyo, Japan)
• CDpro package: CDSSTR, CONTIN/LL and SELCON 3

Circular dichroism spectroscopy (CD) measurements were performed in quartz cuvettes of 0.1 cm path length using a Jasco J-815 spectrometer. Spectra were recorded at 20 °C to 90 °C in steps of 10 °C from 195 to 250 nm with a resolution of 1.0 nm and an acquisition time of 20 nm/min. Final CD spectra were obtained by averaging five consecutive scans and they were corrected for background by subtraction of spectra of protein-free samples recorded under the same conditions. Mean residue ellipticity (Θ) was calculated based on the molar protein concentration and the number of amino residues of the Tom40 isoforms. The protein concentration used for CD spectroscopy was adjusted to 0.2 - 0.5 mg/ml. Sample buffer was used for baseline determination.

The secondary structure content was estimated using the CDpro package according to Sreerama, namely CDSSTR, CONTIN/LL and SELCON 3 (Sreerama and Woody 2000; Sreerama and Woody 2003; Sreerama and Woody 2004). Melting curves were recorded at constant wavelength at 216 nm for recombinant hTom40A/B from 20 to 98 °C by applying a temperature ramp of 1 °C/min. The percentage of unfolded protein content $f_U(T)$ was calculated according to $f_U(T) = (\Theta(T) - \Theta_N(T))/(\Theta_U(T) - \Theta_N(T))$, where $\Theta_U(T) = aT + b$ and $\Theta_N(T) = cT + d$ represent the pre- and post-transition baselines. For evaluating the protein melting temperature $T_m$ the resulting data $f_U(T)$ were fitted by the Boltzmann equation $f_U(T) = (f_0 - f_{max})/(1 + \exp(T - T_m)/T_s) + f_{max}$, where $f_0$, $f_{max}$ and $T_s$ are the minimum and maximum percentages of unfolded protein content and the temperature range over which the transition occurs, respectively.

### 2.5.3 Fourier Transformation Infrared Resonance Spectroscopy (FTIR)

Devices: • TENSOR 27 FTIR spectrometer
(Bruker Optik GmbH, Ettlingen Germany)

Software: • OPUS Quant 2 (Bruker Optik GmbH, Ettlingen Germany)

For FTIR spectral measurements protein samples containing recombinant Tom40 were prepared as follows. The protein solution had a concentration of 5-10 mg/ml and was dialyzed against buffer to obtain a final detergent concentration of 0.1 % LDAO as the protein was most stable in this detergent at high concentrations. FTIR-spectra measurements were performed using a Bruker Optics Confocheck system. The system is based on a TENSOR 27 FTIR spectrometer equipped with a linear, photovoltaic MCT detector. All spectra were recorded for 25 sec with a wave number resolution of 4 $cm^{-1}$. To avoid temperature induced variations of the water signal the measurement cell was kept at a constant temperature of 25 °C using a thermostat. For each spectrum, 32 interferograms were collected and averaged. The aperture setting was 6 mm and the scanner velocity was at 10 kHz. All procedures were carried out to optimize the quality of the spectrum in the amide I region, between 1600 and 1700 $cm^{-1}$.

Calculation of the secondary structure content was done with a multivariant pattern recognition method supplied by Bruker Optics. There, the spectral data are factorized and compared to reference data from proteins of known structure from x-ray crystallography. A library of more than 40 proteins of known structures (source: Protein Data Base, http://www.pdb.org) and concentration measured in water was used to determine the secondary structure of the analyzed protein. The quantitative determination of α-helix and β-sheet were set up using the OPUS Quant 2 software that utilizes the Partial Least Squares (PLS) algorithm which is part of the Confocheck system. The advantages and limitations of such pattern recognition methods in protein FTIR are discussed in Fabian (2000).

### 2.5.4 Electron microscopy

Devices:
- Tecnai $G^2$ Sphera transmission electron microscope (FEI, Eindhoven, Netherlands)
- Tietz F224 CCD camera (TVIPS, Gauting, Germany)

Software:
- Tecnai User Interface TUI (FEI, Eindhoven, Netherlands)
- EM Menu 3.0 (Tietz Camera software, TVIPS, Gauting, Germany)

Recombinant Tom40 was visualized by electron microscopy. Protein in the different detergents LDAO, oPOE, β-OG and Bri35 was diluted to a concentration of 0.1 mg/ml and adsorbed to glow-discharged carbon-coated copper grids (400 mesh, Agar Scientific, Stansted Essex, UK). The grids were washed with dd$H_2O$, blotted with filter paper and stained with 2 % uranyl-acetate solution for 1 min. Images were taken by a FEI transmission electron microscope, equipped with a $LaB_6$ cathode and at a magnitude of 50.000 at acceleration of 200 kV and 2-3 µm underfocus. Digital images were taken with a CCD-camera and processed with the software EM Menu.

### 2.5.5 Fluorescence spectroscopy

Device:
- FP-6500 spectrofluorimeter (Jasco Inc., Tokyo, Japan)

Software:
- Spectra Manager Version 1.53.04 (Jasco Inc., Tokyo, Japan)

Tryptophan fluorescence measurements of recombinant Tom40 were performed at 25 °C on a spectrofluorimeter after dialysis of protein samples in 20 mM Tris-HCl, pH 8, 1 % LDAO, 1 mM β-ME, and 7 M guanidine-hydrochloride. Before measurements were carried out, all samples were diluted 1:20 in buffer containing 0.3 to 7 M guanidine hydrochloride. Samples were incubated at 25 °C for 24 h to ensure thermodynamic equilibrium and spun down at 16.000 x g for 5 min at room temperature. Tryptophanes of recombinant Tom40 were excited at 280 nm in quartz microcuvettes of 0.1 cm path length, and emission spectra were recorded from 300 to 400 nm using an integration time of 1 s. The band widths for excitations and emissions were set to 3 nm with a response of 0.2 s, respectively. Data were corrected by subtracting the appropriate blank value of protein-free sample.

Spectra were analyzed by three different approaches: either a fixed wavelength was set and the signal decrease towards the increase of guanidine hydrochloride was assessed, the shift of the wavelength where the spectra show a maximum or the integral of the emission spectra were plotted against the guanidine hydrochloride concentration. Fluorescence spectra were evaluated by fitting background-corrected spectra $I(\lambda)$ to the log-normal distribution $I(\lambda) = I_0\, e^{-[\ln 2/\ln^2 \rho][\ln^2[1+(\lambda-\lambda_{max})(\rho^2-1)]/\rho\Gamma]}$ (Ladokhin et al. 2000; Winterfeld et al. 2009), where $I_0$ is the fluorescence intensity observed at the wavelength of maximum intensity $\lambda_{max}$, $\rho$ is the line shape asymmetry parameter and $\Gamma$ is the spectral width at half-maximum fluorescence intensity $I_0/2$. To characterize the thermodynamic properties of denaturant induced unfolding of hTom40A the fraction of unfolded protein $f_U(D)$ was fitted to the vant' Hoff equation $f_u(D) = 1/\left(1 + e^{(\Delta G_U^{H_2O} - m[D])/RT}\right)$ where $\Delta G_U^{H_2O}$ is the free energy of the protein that describes its conformational stability at zero denaturant concentration. The factor $m$ is a measure of the dependence of $\Delta G_U$ on denaturant concentration ($D$). In the unfolding transition region it can be described by the linear extrapolation method (LEM) according to $\Delta G_U = \Delta G_U^{H_2O} - m(D)$ (Pace 1986; Myers et al. 1995; Huyghues-Despointes et al. 2001).

### 2.5.6 Laser Induced Liquid Bead Ion Desorption (LILBID)

Devices:
- Droplet generator
  (microdrop Technologies GmbH, Norderstedt, Germany)
- Optical parametric oscillator (custom-made)
- Wiley-McLaren TOF reflectron mass spectrometer (custom-made)

Software:
- Labview (National Instruments, Austin, TX, USA)

All mass spectrometric analyses were carried out by the group of Prof. Bernhard Brutschy from the Institute for Physical and Theoretical Chemistry at the Goethe University, Frankfurt in a cooperation project. The devices have been build and set up by the group of Prof. Brutschy.

The method is described in detail in (Morgner et al. 2007). Briefly, micro droplets of protein solution (diameter 50 µm, 65 pl) were produced on demand at 10 Hz by a piezo-driven droplet generator. The absolute amount of protein in the droplets lied in the femtomolar ($10^{-15}$ M) to attomolar ($10^{-18}$ M) range depending on the concentration. The protein droplets were introduced into vacuum via differential pumping stages where they were irradiated one by one by synchronized high-power mid-IR laser pulses of typically 5 ns pulse length. These were generated in a home-built optical parametric oscillator (OPO) using LiNbO3 crystals and a Nd-Yag laser as pump. The wavelength of the OPO radiation was tuned to the absorption maximum of water at around 3 µm corresponding to an excitation of its stretching vibrations. At threshold intensity the droplet exploded resulting in the emission of ions from liquid into gas phase. There, they were mass analyzed in a time of flight (TOF) reflectron mass spectrometer with a Wiley-McLaren type acceleration region (Wiley and McLaren 1955) and an ion reflectron.

To detect very large biomolecules, a Daly-type ion detector was used, working up to an m/z range in the low Megadalton region ($10^6$ MDa). At low laser intensity LILBID desorbed ions out of the liquid very gently (ultrasoft mode) enabling detection of the noncovalently assembled protein complexes. At higher laser intensities the complex was thermolysed into subcomplexes (soft mode) and further to its covalent subunits (harsh mode). The signals from the detector were recorded by a transient recorder. For data acquisition and analysis a user-written labview program was used. The signal to noise ratio was improved by subtracting an unstructured background, caused by metastable loss of water and buffer molecules, from the original ion spectra. These difference spectra were smoothed by averaging the signal over a pre-set number of channels of the transient recorder, with the smoothing interval always lying within the time resolution of the TOF mass spectrometer. The recorded mass spectra were usually averages of 100-200 droplets which resulted in the consumption of less than 10 µl of protein solution. The spectra were calibrated with solubilized bovine serum albumin (67 kDa).

For laser-induced liquid bead ion desorption mass spectrometry analysis of TOM core complex, the protein solution was concentrated to a final concentration of ~2 mg/ml (~5 µM) using spin concentrators with a molecular mass cutoff of 5 kDa (Millipore GmbH, Germany) and dialyzed at 4 °C against 0.05 % DDM, 50 mM NaCl and 20 mM Tris-HCl, pH 8.5. Finally, the sample was transferred into 30 mM $NH_4HCO_3$ pH 6.8, 0.05 % DDM using Zeba™ Micro Desalt Spin Columns (Pierce, Thermo Fisher Scientific, USA) following standard procedures.

### 2.5.7 Electrophysiology

Devices:
- EPC-8 patch clamp amplifier (HEKA Electronics, Lamprecht, Germany)
- A/D converter NI-USB-6251
  (National instruments, München, Germany)
- Oscilloscope HM 504 (HAMEG Instruments, Mainhausen, Germany)
- Chamber Delrin cup BCH-13a
  (Warner Instruments, Hamden, CT, USA)

Software:
- WinEDR 3.9 (J. Dempster, University of Strathclyde, Glasgow, UK)
- WinWCP 3.6 (J. Dempster, University of Strathclyde, Glasgow, UK)
- pClamp-Suite 8 (Axon Intruments, Union City, CA, USA)

Buffers:
- KCl-buffer: 50 mM HEPES pH 7.2, 1 M KCl
- Priming lipid: 0.5 % DiphPC (Avanti Polar Lipids, Alabaster, AL; USA) in Methanol and Chloroform (1:1)
- Painting lipid: 0.5 % DiphPC (Avanti Polar Lipids, Alabaster, AL; USA) in n-decane and butanol (9:1)

For qualitative analysis of Tom40 channel characteristics purified protein was reconstituted into black lipid membranes (BLM) and channel currents were recorded according to standard protocols (Engelhardt et al. 2007; Poynor et al. 2008). Proteins used for channel measurements were purified as described in 2.4.5.2 and 2.4.5.3. The detergents of choice for the electrophysiological measurements were oPOE and LDAO as they appeared to facilitate channel insertions. Additionally, protein solution was saturated with cholesterol as this is a component in cell membranes of animals and might facilitate the insertion of human channel proteins like hTom40. Priming and painting lipids were prepared by diluting stock solutions of 1 % DiphPC in chloroform in organic solutions using a Hamilton syringe (VWR, Darmstadt, Germany). Lipid dilutions were freshly prepared approximately every five days.

A bilayer chamber from Warner instruments was equipped with cylindrical delrin cup containing an aperture of 250 µm in diameter. To ensure convenient membrane formation the outside of the aperture was treated with 2 µl of priming lipid. When the lipid had completely dried the delrin cup was placed in the chamber and both sides of the chamber were filled symmetrically with KCl-buffer. To obtain membranes a teflon loop (home-built) with 1 µl of painting lipid was swept over the cis-side of the aperture until a membrane was formed. The thickness of the membrane was controlled by determining the membrane capacity. A double layer membrane over the aperture of 250 µm had a capacity of ~300 pF.

Protein solution prepared as described above was added to the cis-side of the membrane in a final concentration of 10-50 µg/ml. Current fluctuations through single channels were recorded using a patch-clamp amplifier in voltage-clamp mode. The headstage of the amplifier was connected to the bilayer chambers by a pair of Ag/AgCl pellet electrodes (WPI, Berlin, Germany). Current signals were low-pass filtered at 3 kHz using the built-in Bessel-filter of the amplifier and monitored for channel insertion using an analogue oscilloscope. Current and voltage signals from the amplifier were digitized at a sampling rate of 10 kHz per channel using a NI-USB-

6251 interface controlled by a program of the Strathclyde electrophysiology suite (WinEDR 2.8 or WinWCP 3.6).

### 2.5.7.1 Single channel analysis

For determination of single channel conductance of recombinant Tom40 a constant voltage of 10 -50 mV was applied until an incorporation of a channel was observed as a step-like current increase. All recorded data points of a current measurement at a steady voltage were divided into bins of a specific value. The frequency of points within the individual bins were normalized to the bin width and plotted against the corresponding bin center. If more than one conductance state was present the histogram revealed more than one peak. All peaks were fitted with multiple Gaussian peak functions and the mean position (current amplitude) was determined. The difference between two current amplitudes was divided by the applied voltage which results in the conductance of the channel defined as $G[nS] = I[pA]/U[mV]$. The conductance $G[nS]$ was plotted against the frequency and accumulation at certain conductivities shows possible conductance states of the channel.

### 2.5.7.2 Analysis of voltage dependence

For analysis of voltage dependence currents were recorded in response to either linear voltage ramps or stepwise voltage changes. Voltages were initially applied for 30sec until the current was stable. In further measurements the voltage application time was reduced to 10 sec if the analyzed channel did not show current decrease in 30 sec. A baseline without membrane potential was always measured in between the stepwise voltage changes and the difference between current and baseline was determined. These data points of current measurements were normalized as described above and the mean value was divided by the applied voltage. A normalized conductivity ($G_{norm}$) was designated as $G_{norm} = (G_{+10mV} + G_{-10mV})/2$. Values for $G_n/G_{norm}$ were plotted against the applied voltage. Variations from $G_n/G_{norm}$ towards zero indicated a loss in channel conductivity and could result in the formation of bell-shaped curve and which was eventually fitted with the double-Boltzman equation in the form $P(V) = 1/[(1+e^{A_l \cdot (V-V_{0l})}) \cdot (1+e^{A_r \cdot (V-V_{0r})})]$ where $V_{0l}$ and $V_{0r}$ are the voltages at which the open probability $P(V)$ is half maximal and $A_l$ and $A_r$ are the voltage sensitivities.

## 2.5.8 3D-Crystallization

Crystallization trials in this work have been performed for TOM core complex, NcTom40 and human Tom40AΔ1-82 and human Tom40AΔ1-82$^{3mut}$. The proteins were purified via Ni-NTA affinity chromatography and size exclusion chromatography as described above. A high protein concentration of > 5 mg/ml was essential to increase the chances for crystallization. After purification the proteins were concentrated with spin columns from Vivaspin or Millipore with a cut-off of 30 kDa. For hTom40AΔ1-82 and hTom40AΔ1-82$^{3mut}$ a protein concentration above 10 mg/ml was achieved. Eventually, the concentrated proteins were dialyzed against their initial purification buffer (see 2.4.5.3 and 2.4.5.4.).

### 2.5.8.1 Crystallization trials at the Max Planck Institute (MPI) in Tübingen

Device: • Honeybee 961 crystallization robot (Formulatrix Inc., MA, USA)

Software: • RockMaker (Formulatrix Inc., MA, USA)
• RockImager 54 (Formulatrix Inc., MA, USA)

Crystallization trials were set up on 96-well Corning 3350 plates (Hampton Research, Aliso Viejo, CA, USA) at the Max-Planck-Institute (MPI) in Tübingen in collaboration with Kerstin Bär or Reinhard Albrecht. The protein concentration for the trials ranged from 5 mg/ml for TOM core complex and 2 mg/ml for NcTom40 to 7 mg/ml for hTom40. For crystallization trials by vapor diffusion 200-500 nl of protein solution was mixed with the same or variable amount of screening buffer in over 1000 different buffer conditions. Crystallization trials were set up according to the sitting drop method next to a buffer reservoir. Crystal growth was monitored by automatic imaging of the 96-well plates after 1, 7 and 30 days with the Rock Imager software. After successful crystal growth, the crystals were picked with a nylon loop, frozen in liquid nitrogen and diffraction was assessed at the synchrotron radiation facility Swiss light source (SLS, Villigen, Switzerland) at the beamline PX10.

| Crystal Screens MPI Tübingen: | |
|---|---|
| Ozma PEG 1K, 4K, 8K, 10K | Emerald Biosystems |
| Cryo I, II | Emerald Biosystems |
| Wizard I, II, III | Emerald Biosystems |
| Index HT | Hampton Research |
| Crystal Screen HT | Hampton Research |
| Salt Rx | Hampton Research |
| MembFac HT | Hampton Research |
| Crystal Screen Cryo | Hampton Research |
| PEG / Ion Screen | Hampton Research |
| JBS Screen Classic HTS I, II | Jena Bioscience |

| | |
|---|---|
| Protein complexes | Sigma-Aldrich |
| The JCSG+ Suite | Quiagen |
| The PACT Suite | Quiagen |

## 2.5.8.2 Crystallization trials at the Structural Biology Institute (IBS) in Grenoble

Devices: • PixSys 4200 crystallization robot
(Genomic Solutions, Huntingdon, U.K.)

Software: • Data Collection Software IsPyB (developed by ESRF, Grenoble, France)
• RockImager (Formulatrix Inc., MA, USA)

Further crystallization trials were set up at the Structural Biology Institute (IBS) in Grenoble in collaboration with Jacques-Phillippe Colletier and coworkers. Classical crystallization was first attempted, i.e. with the protein solubilized in detergent micelles (comparable to 2.5.8.1). Crystallization was also attempted following the solubilization of the protein in "bicelles" or in a "lipidic cubic phase", respectively.

Crystallization trials in bicelles (Faham and Bowie 2002) were performed following the mixing of the protein with either 8 or 32 % bicelles solutions (DMPC/CHAPSO), at a protein/bicelle ratio of 4:1. The initial protein concentration of 10 mg/ml was therefore lowered to 8 mg/ml in 1.6 % or 6.4 % bicelles, respectively. From a practical point of view, crystallization in bicelles is similar to crystallization in detergent micelles, the only requirement is that the protein/bicelles solution is kept at 4° C before being mixed with a given precipitant. Therefore, robotic crystallization was envisaged without any further adapting, and screening performed among various conditions using 100 nl of the protein/bicelles solution at the time. Pipetting was performed by a by Cartesian nanovolume crystallization robot on 96-well plates (Greiner Crystal Quick plates, Hampton Research). In our case, the incubation temperature for crystal growth was also 4 °C, and the hanging-drop vapor diffusion method was used. The buffer screens tested with these methods are listed below. Images of the plates were taken with the RockImager software after 24 h, 72 h, 7 d, 16 d, 5 weeks and 10 weeks and made available online for observation.

**Crystal Screens IBS Grenoble:** Crystal Screen Lite — Hampton Research
Crystal Screen Natrix — Hampton Research
MembFac HT — Hampton Research
Ammonium-Sulfate — Hampton Research
PEG 6K — Hampton Research
Screen Index — Hampton Research
Screen MPD — Hampton Research
The classic Suite — Quiagen
The PEGs — Quiagen
The pH clear — Quiagen
Screen Mme 5000 — HTX-Lab

Crystallization trials were also performed using the lipidic cubic (*meso*) phase (Landau and Rosenbusch 1996). Monoolein was used to yield cubic phases from membrane proteins solution. To crystallize human Tom40A in meso phase, 60% monoolein cubic phases containing either the wild type or mutant protein were generated by emulsification; the initial protein concentration was 14 mg/ml for the wild type and 10 mg/ml for the mutant (i.e. 6.4 mg/ml and 4 mg/ml in the cubic phase, respectively). So, 50 µl of mesophase proteins was composed of 30 µl monoolein and 20 µl of protein. Drops of 100 nl of the resulting "gel"-like cubic phase were deposited on a 96-wells trays to which 2 µl of various precipitant solutions were added. Six different screens each with 96 conditions were tested robotically. The incubation temperature for crystal growth was set at 22 °C. All crystallization trials have been carried out at the high throughput crystallization laboratory (HTX lab) of the EMBL in Grenoble (https://embl.fr/htxlab/).

**Screens lipidic cubic phase for hTom40A wild type:**

MbClass II Suite — Quiagen
CubicPhase II Suite — Quiagen
MemGold — Molecular Dimensions
MemPlus — Molecular Dimensions

**Screens lipidic cubic phase for hTom40A K107L/H117L/H220L:**

CubicPhase I Suite — Quiagen
CubicPhase II Suite — Quiagen
MemStart + MemSys — Molecular Dimensions
MemPlus — Molecular Dimensions

### 2.5.8.3 Diffraction experiments at the European Synchroton Radiation Facility (ESRF) in Grenoble

Software: • Imaging Software Fit2D (developed by ESRF, Grenoble, France)

The crystals qualities were assessed by their diffraction following exposure of the crystal to an x-ray beam. Data shown in this work were collected at the European Synchrotron Radiation Facility (ESRF) in Grenoble at two beamlines, either the ID23EH2 microfocus beamline (5 µm beam; $\lambda$= 0.86 Å; MarCCD detector) or the BM30A beamline (300 µm beam; $\lambda$= 0.97 Å; ADSC-315r CCD detector). Images were produced using the software Fit2D.

To avoid ice formation in the crystals, which would lead to a disruption of crystalline order, mother liquor solutions containing 18 % glycerol were prepared for each crystal type beforehand, in which crystals were transferred before their flashcooling to 100 K for data collection. Sufficiently large crystals were mounted in a nylon loop and flash-cooled to 100 K directly in the nitrogen gas stream (Oxford Cryosystems 700). Crystals, smaller than 5 µm in radius, were mounted in batch on a kapton-grid loop, before flashcooling them directly in the nitrogen gas stream.

# 3 Results

## 3.1 Stoichiometry of TOM core complex

The composition of subunits in the TOM complex represents a key step to gain insight into the structural organization of the important protein machinery in the outer mitochondrial membrane. The arrangement of subunits gives hint about the translocation process which is still not well understood. With classical methods to determine the mass of complexes like size exclusion chromatography or BN-PAGE progress has been made to approach the mass of the complex but a definite mass or even the stoichiometry could not been solved (Kiebler et al. 1990; Dekker et al. 1998; Ahting et al. 1999; Werhahn et al. 2001). With a rather new method combining electrospray-ionization and mass spectrometry developed by the group of Prof. Bernd Brutschy (Goethe University, Frankfurt) great improvement has been achieved in accuracy and sample consumption. The method, named laser-induced liquid bead ion desorption (LILBID), is taking advantage of the ionization of proteins and their eventual detection in a TOF-analyzer. In contrast to electrospray ionization (ESI) LILBID enables ionization at higher salt concentrations and in detergent buffers and favors single charged ionization states of the analytes (Morgner et al. 2006). Several protein complexes revealed their subunit composition when analyzed with LILBID (Hoffmann et al. 2010; Sokolova et al. 2010) and therefore seemed suitable for stoichiometry analysis of the TOM complex.

### 3.1.1 Purification of TOM core complex

TOM complex has been purified in the lab of Prof. Stephan Nußberger for many years according to established protocols (Ahting et al. 1999; Ahting et al. 2001). The organism of choice for the expression of genetically modified TOM complex is the fungus *Neurospora crassa* which is easy to cultivate in large scales. The strain used in this work has a hexahistidinyl-tag at the C-terminus of the secondary receptor Tom22 and is termed GR 107 (Künkele et al. 1998).

A culture of 80 L results in the harvest of ~1.5 -2.5 kg of hyphae after 24 h of growth. Mitochondria isolation from 1 kg of hyphae gives about 3-4 g of mitochondria solubilized at a concentration of 50 mg/ml. Solubilization of mitochondrial membranes with the mild detergent n-dodecyl-β-D-maltoside solubilizes the TOM complex along with all mitochondrial membrane proteins in detergent micelles. The primary receptors Tom20 and Tom70 dissociate from the complex during this procedure. The resulting core complex consisting of Tom40, Tom22 and the small proteins Tom5, Tom6 and Tom7 is stable in detergent solution at pH 8.5.

Purification of TOM core complex via Ni-NTA affinity chromatography elutes the complex along with other histidin-containing proteins from mitochondrial membranes. Further purification via and anion exchange chromatography removes most of these impurities and reveals the TOM complex that contains all known subunits with 99 % purity (Figure 3.1). Tom40 and Tom22 are detected as strong bands on SDS-polyacrylamide gels. The small molecular mass components Tom5, Tom6 and Tom7 do not separate well on small SDS-gels but can be visualized on Tricine-SDS-gels (Figure 3.1). The presence of all subunits was confirmed by Western blotting and specific immunodetection (data not shown).

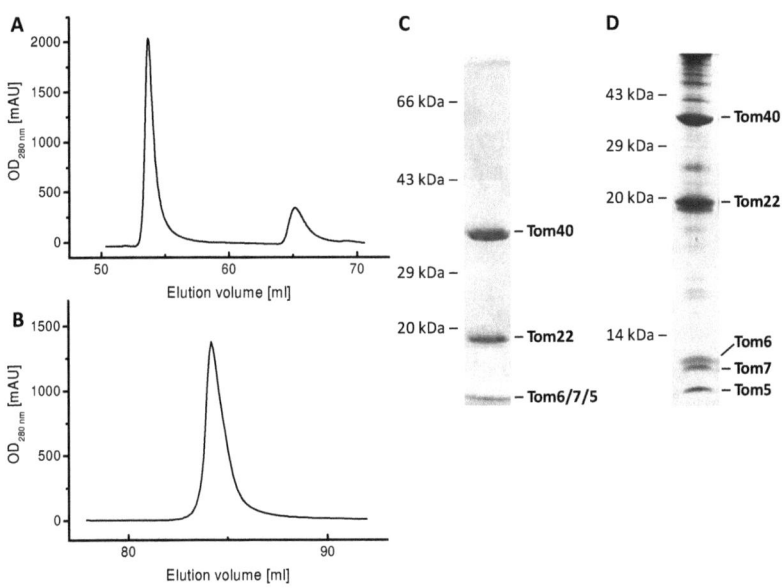

Figure 3.1: Subunit composition of purified TOM core complex of *N. crassa*. (A) Elution profile of *N. crassa* TOM core complex subjected onto Ni-NTA High HisTrap column, eluted at 300 mM imidazole; (B) Elution profile of *N. crassa* TOM core complex from anion exchange chromatography (Resource Q). TOM core complex was eluted in a single peak fraction by a step gradient with 400 mM KCl at pH 8.5. (C) Analysis of the core-complex subunits by SDS-PAGE followed by Coomassie staining, Tom40 and Tom22 can be detected, the small Tom-proteins are not separated (D) Analysis of the core-complex subunits by Tricine SDS-PAGE followed by silver staining reveals all known subunits of the core complex, Tom40, Tom22, Tom7, Tom6 and Tom5 and some impurities.

## 3.1.2 LILBID-mass spectrometry

LILBID mass spectrometry together with electrospray-ionization was used to analyze the quartenary structure of the high-mass protein oligomer TOM. The method implies a "top-down" approach to analyze protein-protein interactions and their assembly. Depending on the applied laser intensity these interactions are still functional or disrupted. This gives broad information about the interaction of different proteins among each other or the formation of oligomers of the same protein. The eventual assignment of proteins to certain peaks in the mass spectrum is only possible when the respective protein masses in the analyte are known.

The mass spectrum of purified TOM core complex (Figure 3.2) recorded at a high laser intensity, where membrane protein complexes fully disintegrate into their subunits (Morgner et al. 2007), shows a clear set of peaks over the mass range from 0 to 42,000 m $z^{-1}$. Most peaks can readily be assigned to the five constituents of the TOM core complex, i.e. Tom40, Tom22 and the small Tom proteins Tom7, Tom6 and Tom5 (Table 8) which underlines data from SDS-PAGE (Figure 3.2). Only three small peaks marked at 8,400 m $z^{-1}$, 9,000 m $z^{-1}$ and 9,900 m $z^{-1}$ were of unknown origin which might be excited from impurities in the sample (Figure 3.1).

Comparison with the theoretical masses of the Tom subunits by sequence analysis (ExPASy, Swiss Institute of Bioinformatics) indicates that all detergent molecules were stripped off. This is of advantage for the correct assignment of peaks to the according protein masses. This fact also gives the method an advance towards other mass determining approaches where the detergent micelle usually still surrounds the protein and therefore falsifies the mass determination. Interestingly, a strong signal was also visible at 24,102 m $z^{-1}$ indicating a tight association of Tom22 (17,809 m $z^{-1}$) with most likely Tom6 (6,407 m $z^{-1}$). This result stands in in line with biochemical studies addressing the stability of TOM under conditions where the complex was disintegrated by non-ionic detergents with short alkyl-chains where a tight interaction between Tom22 and Tom6 had been observed (Ahting et al. 2001; Dembowski et al. 2001). At low m $z^{-1}$ values the resolution of LILBID is below 1 kDa and in this range the correct assignment of the small Tom proteins is possible (Figure 3.2). The interaction between Tom22 and Tom6 shows a unique constellation and does not occur with the other two small Tom proteins Tom5 and Tom7. It is remarkable that there are no peaks identified which would correspond to the mass of two small Tom proteins implying that a binding between them does not occur in the complex or is too weak to be identified.

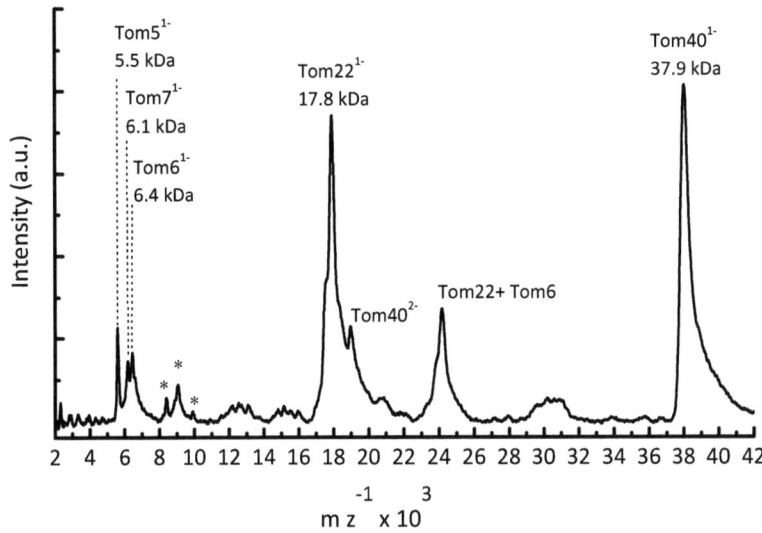

Figure 3.2: LILBID mass anion spectrum of TOM core complex using high laser intensity. The most intense peaks were assigned to the five constituents of the translocation core complex, Tom40, Tom22, Tom7, Tom6 and Tom5 are visible as singly charged molecules. The peak at 18,962 m $z^{-1}$ represents Tom40 molecules with two negative charges. The peak at 24,102 m $z^{-1}$ can be assigned to one Tom6 subunit associated with one Tom22 molecule. Peaks marked with * are of unknown origin. The theoretical masses of the proteins (ExPASy) are assigned to the peaks.

To investigate the stoichiometric composition of the TOM core complex, spectra of TOM were recorded under soft desorption conditions where the complex disintegrates into subcomplexes and individual subunits (Figure 3.3, Table 8). Strong intensities were observed for Tom22 and Tom40 in their monomeric form in agreement with Figure 3.2 as well as Tom22 and Tom40 associated with one or two small Tom subunits, respectively. Smaller peaks were identified as complexes composed of two Tom40, two Tom22 and several small Tom molecules. Due to the resolution limit of LILBID-MS at high m $z^{-1}$ the exact identity of the small Tom proteins between 5.5 and 6.4 kDa associated with Tom22 and Tom40 could not be determined with certainty. This again gives hint about the strong interaction between Tom20 and also Tom40 to the small Tom proteins which has been addressed above. Remarkably, Tom22 is not only capable of binding one Tom6, but possibly even two small Tom proteins, as indicated by the broad peak centered at 24,000 m $z^{-1}$. The broadness of the peak could imply the variability concerning the binding partners Tom5, Tom7 or another Tom6. However, a tight binding between Tom22 and a small Tom protein has previously only been reported for Tom6 (Dekker et al. 1998; Ahting et al. 2001; Dembowski et al. 2001).

A stated above, is the resolution of LILBID at medium laser intensities at around 1 kDa. However, this is not sufficient to differentiate between the mass of 1 x Tom22 (17.8 m $z^{-1}$) or 3 x smTom proteins (~ 18 m $z^{-1}$). So it is not clear whether the respective peak at 55.7 m $z^{-1}$ corresponds to 1 x Tom40 + 1 x Tom22 or to 1 x Tom40 + 3 x smTom (Figure 3.3). Nevertheless, the peak at 55.7 m $z^{-1}$ is quite sharp indicating a subcomplex consisting of 1 x Tom40 + 1 x Tom22 rather than 1 x Tom40 + 3 x smTom. Additionally, it is questionable whether Tom40 can bind three or more small Tom proteins alone or only in association with Tom22. The peaks in the spectrum become less intense and smear with higher mass hindering a definite mass determination. Still it becomes clear that the peak difference at higher m $z^{-1}$ values lies in a range of the 6,000 m $z^{-1}$, each pointing to the addition of one small Tom protein.

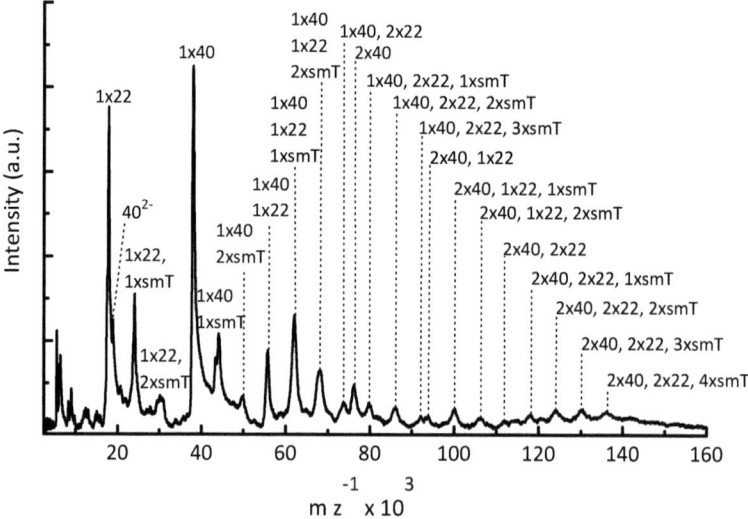

Figure 3.3: Mass spectra of TOM core complex recorded under desorption conditions where the complex disintegrates into subcomplexes and the individual subunits. The spectrum shows multiples of Tom22 and Tom40 as well as complexes with up to two Tom40, two Tom22 and four small Tom molecules marked smT (Tom5, Tom6 or Tom7) with average masses of 6.0 kDa, respectively. All peaks represent singly charged Tom molecules or sub-complexes if not specifically specified (e.g. Tom40$^{2-}$). The peaks at m $z^{-1}$ < 15,000 can be assigned to individual Tom subunits as described in Figure 3.2.

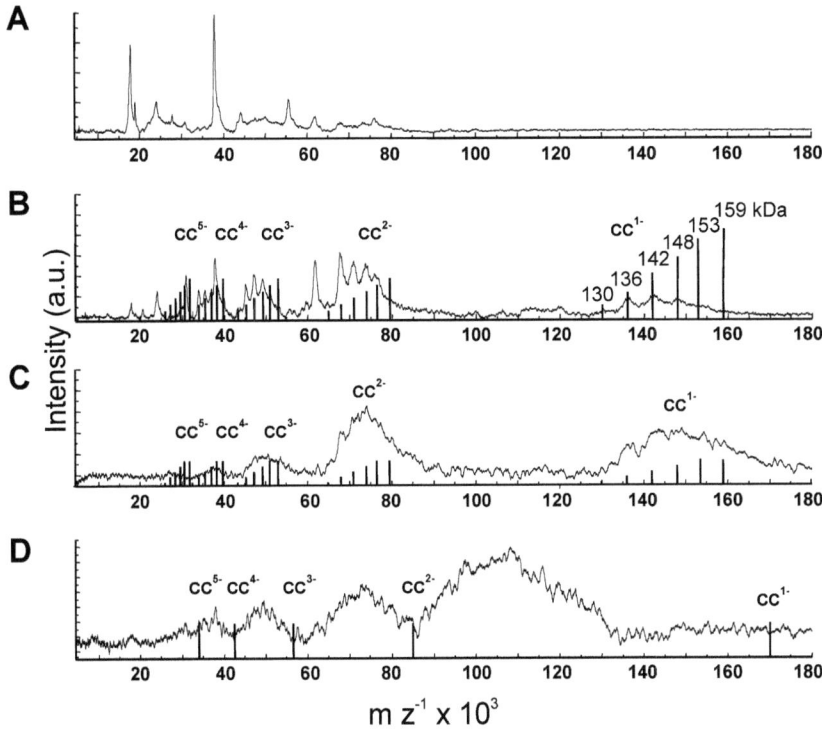

Figure 3.4: Mass spectra of TOM core complex (CC) recorded under decreasing laser intensity. (A) At harsher conditions spectra similar to that in Figure 3.2 appear. At low laser intensity (B) additional series of signals are visible between 25,000 and 170,000 m $z^{-1}$, where z ranges between -5 and -1. Upon lowering the laser intensity further (C) the peaks of the subunits and small subcomplexes disappear and signal peaks corresponding to the multiply charged subcomplexes of a size between 130 and 160 kDa gets more intense due to a reduced charging of the complex. At the same time the peaks are no longer resolved mostly due to reduced signal intensity. It represents a stable complex of 2 x Tom40, 2 x Tom22 and a variable amount of small Tom proteins. At the lowest intensity, when still desorption is possible (D) only broad peaks appear corresponding to a charge distribution of mass 170 kDa.

Decreasing the laser intensity step by step from the harsher conditions (Figure 3.4) as in Figure 3.3 to softer ones (Figure 3.4 B), additional LILBID-MS signals appeared corresponding in a series of protein complexes defining the core complex (CC) with charges between -5 and -1. Upon further lowering the laser intensity (Figure 3.4 C) the peaks corresponding to singly charged molecules could be detected between 130,000 and 160,000 m $z^{-1}$. They are increased in relative intensity but are now unresolved due to decreased signal intensity as detergent and solvent molecules are still attached under these soft conditions. Decreasing further the laser intensity (Figure 3.4 D) the entire mass of the molecule moves towards higher masses of 160-170 kDa.

Table 8: Theoretical and experimental masses of *N. crassa* TOM core complex, TOM subcomplexes and single subunits.

| Protein | Predicted $M_r$[#] (kDa) | LILBID $M_r$ (kDa) | LILBID Charge (z) |
|---|---|---|---|
| Tom40 | 38.151 | 37.9 ± 0.5 | 1-, 2-[a] |
| Tom22-6His | 17.639 | 17.8 ± 0.5 | 1-[a] |
| Tom7 | 6.061 | 6.1 ± 0.1 | 1-[a] |
| Tom6 | 6.463 | 6.4 ± 0.1 | 1-[a] |
| Tom5 | 5.402 | 5.5 ± 0.1 | 1-[a] |
| 2Tom40-2Tom22-6Tom5/6/7 | 147.580 | 148 ± 1 | 1-, 2-, 3-, 4-, 5-[c] |
| 2Tom40-2Tom22-5Tom5/6/7 | 141.580 | 142 ± 1 | 1-, 2-, 3-, 4-, 5-[c] |
| 2Tom40-2Tom22-4Tom5/6/7 | 135.580 | 136 ± 1 | 1-, 2-, 3-, 4-, 5-[c] |
| 2Tom40-2Tom22-3Tom5/6/7 | 129.580 | 130 ± 1 | 1-, 2-[b] |
| 2Tom40-2Tom22-2Tom5/6/7 | 123.580 | 124 ± 1 | 1-, 2-[b] |
| 2Tom40-2Tom22-1Tom5/6/7 | 117.580 | 118 ± 1 | 1-, 2-[b] |
| 2Tom40-2Tom22 | 111.580 | 112 ± 1 | 1-[b] |
| 2Tom40-1Tom22-2Tom5/6/7 | 105.941 | 106 ± 1 | 1-[b] |
| 2Tom40-1Tom22-1Tom5/6/7 | 99.941 | 100 ± 1 | 1-[b] |
| 2Tom40-1Tom22 | 93.941 | 94 ± 1 | 1-[b] |
| 2Tom40 | 76.302 | 76.2 ± 0.5 | 1-[b] |
| 1Tom40-2Tom22-3Tom5/6/7 | 91.429 | 92 ± 1 | 1-[b] |
| 1Tom40-2Tom22-2Tom5/6/7 | 85.429 | 86 ± 1 | 1-[b] |
| 1Tom40-2Tom22-1Tom5/6/7 | 79.429 | 79.8 ± 0.5 | 1-[b] |
| 1Tom40-2Tom22 | 73.429 | 73.7 ± 0.5 | 1-[b] |
| 1Tom40-1Tom22-2Tom5/6/7 | 67.790 | 68.0 ± 0.5 | 1-[b] |
| 1Tom40-1Tom22-1Tom5/6/7 | 61.790 | 62.0 ± 0.5 | 1-[b] |
| 1Tom40-1Tom22 | 55.790 | 55.7 ± 0.5 | 1-[b] |
| 1Tom40- -2Tom5/6/7 | 50.151 | 49.8 ± 0.5 | 1-[b] |
| 1Tom40- -1Tom5/6/7 | 44.151 | 44.0 ± 0.5 | 1-[b] |
| 1Tom22-1Tom5/6/7 | 23.839 | 24.0 ± 0.5 | 1-[b] |
| 1Tom22-1Tom6 | 24.102 | 24.2 ± 0.5 | 1-[a] |

The predicted mass of Tom5/6/7 is 6.0 kDa corresponding to the most frequent difference between the measured peaks of molecules including the small Toms. #: Predicted average molecular mass using ExPASy (Swiss Institute of Bioinformatics); a: Figure 3.2; b: Figure 3.3; c: Figure 3.4 C.

At the lowest laser intensity (Figure 3.4 D) still providing ion desorption, broad peaks appear. They correspond to a complex of 170 ±10 kDa in different charge states (N = 2-5). Here, the width of the peaks can be assigned to detergent molecules which might still be attached to the complex at low laser intensities or possibly also to a few loosely bound small Tom proteins.

Previous chemical cross-linking experiments did not reveal direct contact between Tom40 and Tom22 molecules suggesting that other subunits of the TOM complex act as linking components between Tom40 and Tom22 (Dembowski et al. 2001). The LILBID-MS data shown here clearly indicate that this is not the case. Thus, dimeric Tom40 appears to be the central structural element of the translocation machinery. Dimeric Tom40 tightly binds one or two Tom22 molecules as well as several small Tom proteins. Moreover, LILBID-MS provides direct evidence for a tight non-covalent interaction between two Tom40 molecules in the TOM complex. Our spectra clearly show a stable assembly of Tom40 as dimer. Interestingly, dimer formation does not require the presence of Tom22 or any other Tom subunit.

### 3.1.3 Structural characterization of NcTom40

The stoichiometry and subunit composition of the TOM core complex has been investigated successfully with LILBID. Further structural interest focuses on the translocation pore Tom40. It has been shown by LILBID that the protein forms dimers in the complex but how this dimerization takes place and which amino acids are responsible for this formation remained unclear. This matter was addressed by isolating Tom40 from the complex for structural analysis with 3D crystallization trials.

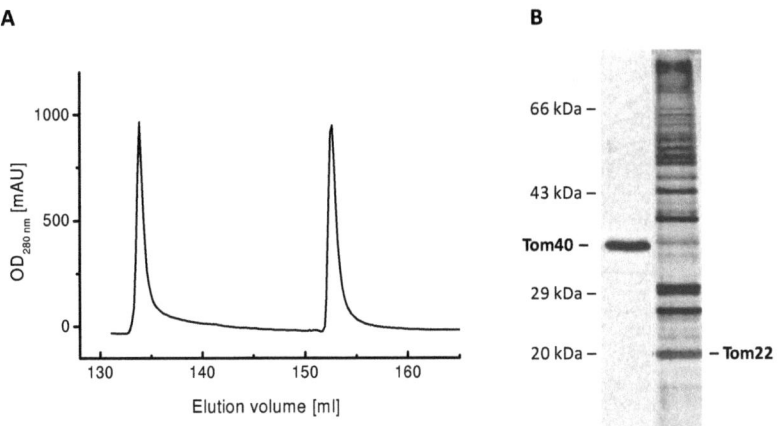

Figure 3.5: (A) Elution profile of *N. crassa* Tom40 and Tom20 purified with Ni-NTA affinity chromatography. NcTom40 was eluted in a single peak fraction with β-OG containing buffer. Tom22 and the residual complex were eluted with 300 mM imidazole. (B) Analysis of NcTom40 (left) and the residual core-complex containing Tom22 by SDS-PAGE followed by silver staining, Tom22 drags along several other outer membrane proteins which can be separated by anion exchange chromatography (data not shown).

Previous protocols have already reported on the purification of Tom40 from the solubilized core complex of *N. crassa* (Ahting et al. 2001). The purification of NcTom40 by the detachment of the protein from the complex with the detergent β-OG has been performed with great efficiency in this work as well. The protein eluted in pure fractions from the Ni-NTA column excluding a second purification step by anion exchange chromatography. The successful purification of NcTom40 has been evaluated by SDS-PAGE and silver staining (Figure 3.5). The yield of NcTom40 eluted from the complex ranged around 0.5 mg per g mitochondria and was significantly lower than the yield of isolated TOM core complex.

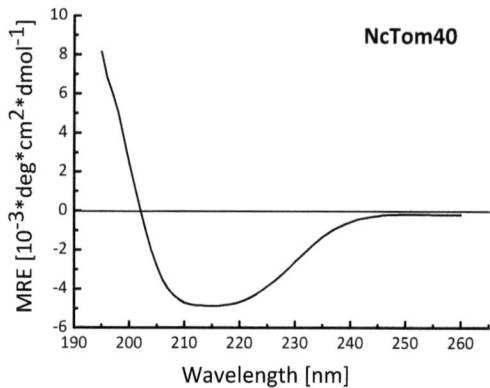

Figure 3.6: CD spectrum of NcTom40 (0.2 mg/ml) in 20 mM $KPO_4$ pH 8, 1 % β-OG hTom40A (~0.2 mg/ml) of accumulated 5 scans at 20 °C and background corrected. Mean residue ellipticity (MRE) was calculated based on the molar protein concentration and the number of amino acid residues of the according protein. The spectrum of NcTom40 indicates a high ratio of β-sheet with a minimum at 212 nm and a crossover of the baseline at 203 nm (with courtesy from A. Schlösinger).

The protein has been dialyzed to suite secondary structure determination. In preparation for crystallization the secondary structure content and thermal stability of NcTom40 was analysed by CD spectroscopy. The spectra show a clear dominance of β-sheet content (Figure 3.6; Table 11). The minimum of the spectra consistently clustered around 212 nm with a crossover of the baseline at 203 nm. At wavelengths > 245 nm, the CD spectrum approached ellipticity values close to zero, indicating that the Tom40 eluted from the complex was virtually free of any higher order aggregates which would cause light scattering effects and interfere with the interpretation of the data. Analysis of data points via CDpro revealed a predominance of β-sheet in the secondary structure (> 40 %) for NcTom40 with low α-helical content (< 5 %). The thermal unfolding of NcTom40 was not evaluated precisely as the protein unfolded slowly over the temperature increase from 20 °C to 95 °C and a definite melting temperature could not been identified.

For an accurate determination of the secondary structure FTIR measurements were carried out for NcTom40. This was problematic as a high protein concentration for NcTom40 was only achievable by concentrating the protein with spin-columns. This leads not only to a concentration of the protein but also of other buffer components including the detergent. Even intensive dialysis of the sample for several days could not adjust the buffer conditions in the sample equally to the reference buffer. The resulting IR-spectra eventually showed a negative signal in the amide I region between a wavenumber of 1700 – 1600 cm$^{-1}$, representing the region in which α-helical, β-sheet and random coil structures are activated This indicated an increased detergent concentration in the protein sample towards reference spectra and prevents a correct secondary structure determination.

NcTom40

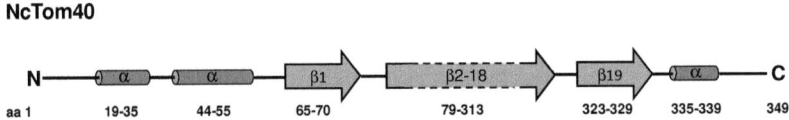

Figure 3.7: Secondary structure model of NcTom40 according to Psipred secondary structure prediction with two helices at the N-terminus, 19 β-strands and one C-terminal helix.

According to Psipred secondary structure determination (Jones 1999; McGuffin et al. 2000) a linear model of NcTom40 has been drawn (Figure 3.7). The model shows the amino acid sequence with the N-terminal segment with two α-helices. The prediction indicates 19 β-strands presumably forming the barrel part followed by a third α-helix at the C-terminus which occurs exclusively in fungal Tom40.

To gain insight into the structure of Tom40 crystallization trials with purified *N. crassa* Tom40 have been set up. The concentration of NcTom40 after the elution from the complex was usually below 1 mg/ml. By concentrating the protein through spin-columns a concentration of 3 mg/ml has been obtained which was suitable for crystallization trials. Crystals grown from these trials were rare and very fragile which excluded further analysis with x-ray radiation (Figure 3.8).

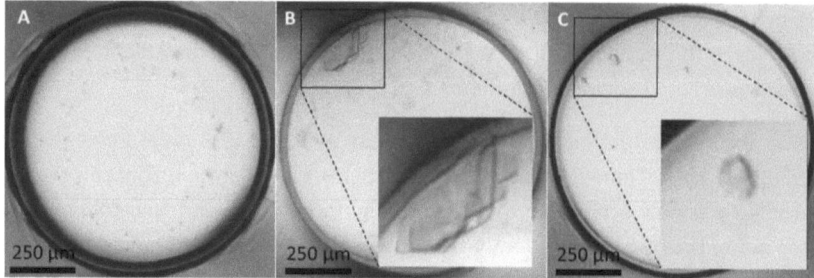

Figure 3.8: Crystals obtained from trials set up with purified NcTom40. Purified NcTom40 in 20 mM HEPES pH 8.5, 1 % β-OG at a concentration of 3 mg/ml. Precipitate and formation of small crystals was observed in conditions with (A) 200 mM NaCl, 100 mM $NaC_2H_3O_2$ pH 6, 30 % /v/v) $C_6H_{14}O_2$, (B) 30 % (w/v) PEG1000, 200 mM $K_2SO_4$ and (C) 140 mM tri-Na-Citrate dehydrate, 70 mM Na-Cacodylate pH 6.5, 21 % (w/v) iso-propanole, 30 5 (v/v) glycerole.

As TOM core complex has been purified in higher concentration of ~5 mg/ml crystallization trials for the complex have been set up as well. However, these trials showed mostly no crystal growth besides precipitate or the formation of very small crystals or needles which were impossible to pick. The heterogenous subunit compositions of TOM core complex, evaluated with LILBID, might explain the difficulties in crystal formation of large protein complexes with a variable amount of subunits as this inhibits the formation of homogenous unit cells in a crystal.

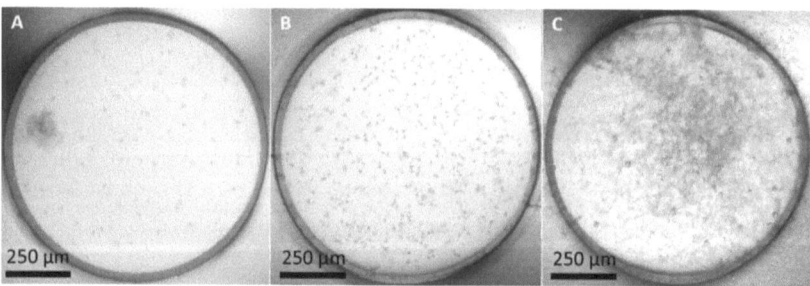

Figure 3.9: Crystals obtained from trials set up with purified N. crassa TOM core complex. Purified TOM core complex in 20 mM HEPES pH 8.5, 0.03 % DDM at a concentration of 5 mg/ml. Needle and precipitate was observed in conditions with (A) 20 % (w/v) PEG 3350, 20 mM NH4, (B) 25 % (w/v) PEG 1500 and (C) 30 % (w/v) PEG 400, 100 mM Na-HEPES pH 7.5, 200 mM $MgCl_2(H_2O)_6$.

## 3.2 Human Tom40

The structure of human VDAC has been solved in the beginning of this project in several groups by x-ray crystallography and NMR (Bayrhuber et al. 2008; Hiller et al. 2008; Ujwal et al. 2008). Due to this fact and the predicted structural relation between human Tom40 and human VDAC1, it was decided to take human Tom40 for structural analysis and crystallization. As the purification of native Tom40 from isolated TOM complex of *N. crassa* did not result in sufficient amounts of pure protein for crystallization human Tom40 was expressed recombinantly in *E. coli*. Human Tom40 occurs in two isoforms, A and B, which differ mainly by their N-terminus. Human Tom40A has an elongated N-terminus of more than 80 residues which is highly enriched in prolines. Human Tom40B has a shorter N-terminus of only 30 residues and not as proline-rich as human Tom40A (Figure 3.10). The possible functions of the N-terminus in human Tom40A are discussed in chapter 4.4. To identify possible structural or functional differences between the two human Tom40 isoforms both proteins were expressed recombinantly.

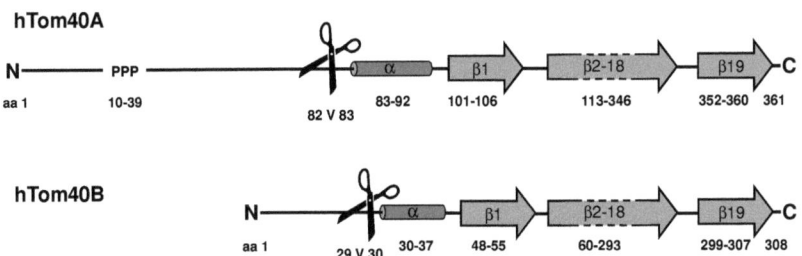

Figure 3.10: PSIPRED-predicted secondary structure of full length hTom40A and hTom40B. Human Tom40A and B were truncated by 82 and 29 amino acids, respectively.

To optimize the structural similarity of human Tom40 to human VDAC1 both Tom40 isoforms were cloned as N-terminally truncated proteins (Figure 1.3). Since the N-termini of both proteins are predicted to be largely disordered and have rather functional than structural requirement it was rationalized that they were not essential for refolding. The published VDAC1 structures further supported the construction of truncated Tom40 proteins assuming that the structural basis of the barrel is shared by all Tom40 proteins (Zeth and Thein 2010) and the N-terminus harbours the most variable part which is not relevant for barrel formation (Figure 1.3). Human Tom40A was truncated by 82 amino acids and human Tom40B by 29 amino acids for structural and functional investigations of the β-barrel domain.

73

### 3.2.1 Expression, purification and refolding of human Tom40

To obtain large amounts of human Tom40 for biochemical and structural studies, hexahistidinyl-tagged human Tom40AΔ1-82 and Tom40BΔ1-29 proteins were expressed in *E. coli* (see Appendix). As no export signal was cloned to the proteins they were incorporated in inclusion bodies. Bacterial cells grew to a cell mass of 5 g per litre of culture. The wet yield of inclusion bodies ranged between 2 and 3 g per litre of culture for both isoforms.

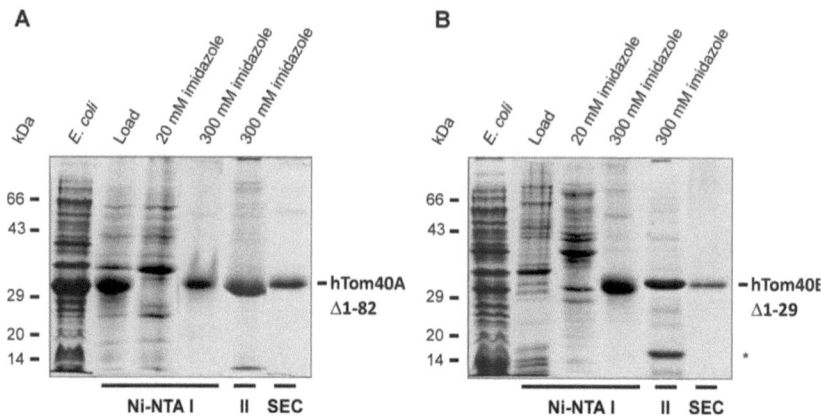

Figure 3.11: Purification of human Tom40AΔ1-82 and Tom40BΔ1-29.
Inclusion bodies isolated from *E. coli* containing recombinant hTom40AΔ1-82 and hTom40BΔ1-29 with C-terminal hexahistidinyl-tags were solubilized by 6 M guanidine hydrochloride and loaded onto a Ni-NTA affinity column. hTom40A and B were eluted under denaturing conditions with 300 M imidazole (Ni-NTA I) and refolded by rapid dilution into 0.5 % LDAO. Refolded proteins were subjected to Ni-NTA chromatography (Ni-NTA II) and eventually passed over a Superose12 size-exclusion column (SEC). Aliquots of the resulting column fractions were analyzed by SDS-PAGE and Coomassie staining. Lane 1: *E. coli* cells expressing hTom40A or hTom40B; lane 2: purified inclusion bodies containing hTom40A or B; lanes 2 to 5: Ni-NTA column fractions and imidazole eluates; lane 6: peak fraction of SEC.

Human Tom40AΔ1-82 was already highly enriched in inclusion bodies (Figure 3.11, A). The first purification with Ni-NTA chromatography of protein in chaotropic buffer containing guanidine-hydrochloride revealed a rather pure protein. The yield of unfolded hTom40AΔ1-82 from one litre of culture was ~200 mg. The expression of human Tom40BΔ1-29 was at lower level at the beginning of this project. Furthermore, a contaminating band was visible in the purified protein fractions on

SDS-PAGE (Figure 3.11, B). This has been identified as a degradation product of human Tom40B by mass spectrometry (data not shown). However, this problem was solved during this project and hTom40B could be efficiently purified as hTom40A. Its yield ranged between 100-200 mg per litre of culture.

A systematic approach was taken to test the ability of recombinant human Tom40 to refold in various detergents. To evaluate optimal refolding conditions a screen over six different detergents at four different pHs has been performed. For refolding, human Tom40AΔ1-82 and Tom40BΔ1-29 proteins (~ 5 mg/ml) in 6 M guanidine hydrochloride were diluted dropwise ten- to twentyfold in different refolding buffer solutions at 4 °C. After 1, 24 and 170 h the efficiency of protein folding was assessed by evaluation of the amount of precipitated protein after centrifugation and determination of the protein concentration in the supernatant by UV absorption at a wavelength of 280 nm. The buffer substances used were 50 mM citric acid pH 5, 50 mM sodium-phosphate-buffer pH 6, 50 mM Tris-HCl pH 7 and 50 mM glycine-sodium-hydroxide pH 10 and contained 1 mM β-mercaptoethanol. The detergents were either n-dodecyl-β-maltoside (DDM), lauryl-dimethylamine-oxide (LDAO), n-octyl-polyoxyethylene (oPOE), Brij35, CHAPS and n-octyl-β-D-glucopyranoside (β-OG) at a concentration of 5 x CMC (Table 9, Table 10).

Table 9: Yield of soluble hTom40AΔ1-82 after refolding

| Detergent | hTom40A 1 h | | | | hTom40A 24 h | | | | hTom40A 170 h | | | |
|---|---|---|---|---|---|---|---|---|---|---|---|---|
| | pH 6.0 | pH 7.0 | pH 8.0 | pH 10.0 | pH 6.0 | pH 7.0 | pH 8.0 | pH 10.0 | pH 6.0 | pH 7.0 | pH 8.0 | pH 10.0 |
| 0.05 % DDM | 75 | 45 | 95 | 100 | 100 | 40 | 45 | 100 | 60 | 25 | 50 | 60 |
| 0.1 % LDAO | 100 | 100 | 100 | 100 | 100 | 100 | 100 | 100 | 100 | 85 | 70 | 90 |
| 0.5 % oPOE | 80 | ≤ 5 | 40 | 95 | 65 | ≤ 5 | ≤ 5 | 35 | 25 | ≤ 5 | ≤ 5 | 15 |
| 0.05 % Brij35 | 75 | 100 | 90 | 100 | 100 | 100 | 85 | 100 | 95 | 100 | 70 | 85 |
| 3 % CHAPS | 40 | 20 | 10 | 40 | 10 | ≤ 5 | ≤ 5 | 15 | ≤ 5 | ≤ 5 | ≤ 5 | ≤ 5 |
| 1 % β-OG | 35 | 15 | 10 | 20 | 25 | ≤ 5 | 10 | 20 | 15 | ≤ 5 | ≤ 5 | ≤ 5 |

hTom40A and hTom40B (Table 10) were refolded in the presence of the detergents DDM, LDAO, oPOE, Brij35, CHAPS and β-OG at different pH. The amount of folded protein was estimated after 1 h, 1 day and 1 week after centrifugation and determination of protein concentration in the soluble phase by UV spectroscopy at 280 nm. The detergent concentrations are given in w/v, the protein concentration is given in % of total.

Table 10: Yield of soluble hTom40BΔ1-29 after refolding

| Detergent | hTom40B 1 h | | | | hTom40B 24 h | | | | hTom40B 170 h | | | |
|---|---|---|---|---|---|---|---|---|---|---|---|---|
| | pH 6.0 | pH 7.0 | pH 8.0 | pH 10.0 | pH 6.0 | pH 7.0 | pH 8.0 | pH 10.0 | pH 6.0 | pH 7.0 | pH 8.0 | pH 10.0 |
| 0.05 % DDM | 25 | 15 | 10 | 25 | 15 | 10 | 25 | 15 | 10 | 25 | 15 | 10 |
| 0.1 % LDAO | ≤ 5 | 30 | 10 | ≤ 5 | 30 | 10 | ≤ 5 | 30 | 10 | ≤ 5 | 30 | 10 |
| 0.5 % oPOE | ≤ 5 | ≤ 5 | ≤ 5 | ≤ 5 | ≤ 5 | ≤ 5 | ≤ 5 | ≤ 5 | ≤ 5 | ≤ 5 | ≤ 5 | ≤ 5 |
| 0.05 % Brij35 | 90 | 60 | 100 | 90 | 60 | 100 | 90 | 60 | 100 | 90 | 60 | 100 |
| 3 % CHAPS | ≤ 5 | 10 | ≤ 5 | ≤ 5 | 10 | ≤ 5 | ≤ 5 | 10 | ≤ 5 | ≤ 5 | 10 | ≤ 5 |
| 1 % β-OG | ≤ 5 | ≤ 5 | ≤ 5 | ≤ 5 | ≤ 5 | ≤ 5 | ≤ 5 | ≤ 5 | ≤ 5 | ≤ 5 | ≤ 5 | ≤ 5 |

The detergent concentrations are given in w/v, the protein concentration is given in % of total.

Optimal refolding of human Tom40AΔ1-82 was achieved by a tenfold dilution of 5 mg/ml protein in 6 M guanidine hydrochloride into LDAO containing buffer at pH 8 (Table 9). No or little refolded protein was observed at acidic pH below 7 or in the presence of detergents such as DDM, oPOE, or β-OG (Table 9). The refolding for human Tom40BΔ1-29 was not as successful as for the isoform A (Table 10). In the refolding screen hTom40B showed good solubility in 0.05 % Brij35 and only moderate solubility in LDAO. However, in following experiments hTom40B showed comparable refolding results as hTom40AΔ1-82, so the same refolding procedure was applied for both isoforms for comparable results.

For further use the LDAO-concentration was reduced to 0.5 % by dropwise addition of the same buffer without detergent. To concentrate refolded human Tom40 the samples were again applied to Ni-NTA affinity. Nonspecifically bound proteins have been efficiently removed with low imidazole concentrations. All fractions containing Tom40 were merged and concentrated up to 10 mg/ml with spin-columns. The concentrations of hTom40AΔ1-82 and hTom40BΔ1-29 were determined by UV absorbance spectroscopy.

Size exclusion chromatography (Figure 3.11) was used for further purification which was necessary for experiments like CD spectroscopy or electrophysiology. Residual amounts of contaminants were removed with size exclusion chromatography (SEC) using a Superose 12 column. After SEC the protein was virtually pure. It is remarkable that both human Tom40A and B eluted from the SEC column in the void volume which would correspond to aggregated protein. To determine the polydispersity of the proteins the samples were analysed with dynamic light scattering (see 3.2.2.1). All samples were filtered and centrifuged to remove possible protein aggregates prior to measurements. The amounts of both proteins obtained were sufficient for electrophysiology, secondary structure determination experiments like CD and FTIR spectroscopy and first crystallization trials for hTom40A.

### 3.2.2 Structural characterization

#### 3.2.2.1 Dynamic light scattering of hTom40

From SEC results, it was assumed that recombinant Tom40 might be aggregated. A second approach to determine particle size in a protein solution was done with Dynamic Light scattering (DLS). DLS measures the scattered light of a sample in very short time periods and correlates the data. This method analyses the distribution of particles of a certain size in solution. The size distribution of hTom40AΔ1-82 in detergent solution showed small particles with a diameter of 4-6 nm (data not shown). The particle size could not define the oligomerization state of the protein but revealed particles of small size and excluded large particles corresponding to aggregates. However, the resolution was not precise enough to differentiate between monomers or lower oligomers. Human Tom40A in different detergent solution of oPOE, β-OG or LDAO resulted in similar particle size distribution. From these results, the protein preparation was determined to be sufficiently monodisperse for crystallization.

#### 3.2.2.2 Secondary structure determination by CD spectroscopy

Efficient refolding of recombinant Tom40 was monitored with CD spectroscopy and the resulting spectra were the first ones shown for human Tom40. The secondary structure content of refolded human Tom40 isoform A and B at a concentration of 0.2 mg/ml in buffer containing 0.1 % LDAO was analysed by CD spectroscopy and compared to spectra of β-barrel proteins with known structure. Spectra of both isoforms show a clear dominance of β-sheet content (Figure 3.12; Table 11). The minimum of the spectra consistently clustered around 216 nm with a crossover of the baseline at 203 nm. At wavelengths > 245 nm, the CD spectra approached ellipticity values close to zero, indicating that the hTom40 preparations were virtually free of any higher order aggregates which would cause light scattering effects and interfere with the interpretation of the data. Deconvolution of the data points with CDpro (Sreerama and Woody 2000; 2003; 2004) revealed a predominance of β-sheet secondary structure (> 30 %) for both human Tom40s with low α-helical content (< 24 %, Table 11). In addition, hTom40A and B exhibited similar CD spectral characteristics with slightly left-shifted spectra for hTom40B. Hence, efficient refolding of both hTom40s could be assumed under the conditions chosen.

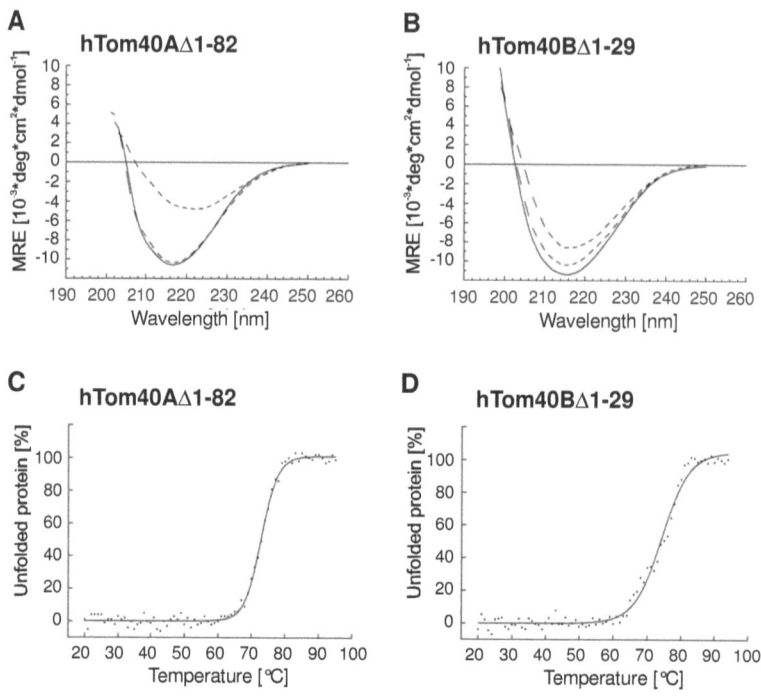

Figure 3.12: CD spectra and thermal stability of hTom40AΔ1-82 and hTom40BΔ1-29

(A and B) hTom40A (~0.2 mg/ml) and hTom40B (~0.2 mg/ml) were solubilized in 0.5 % LDAO, 20 mM Tris, pH 8.0. For each CD spectrum, 5 scans were accumulated at 20 °C (solid line), 50 °C (dashed line) and 70 °C (dashed line) and background corrected. Noisy data below 200 nm have been removed. Mean residue ellipticity (MRE) was calculated based on the molar protein concentration and the number of amino acid residues of the according protein. The spectra of hTom40A and B indicate a high ratio of β-sheet with a minimum at 216 nm and a crossover of the baseline at 203 nm. (C and D) Melting curves of hTom40A and hTom40B determined from corrected CD signals measured at a wavelength of 216 nm. The melting temperatures of hTom40A and hTom40B were at 73 and 74 °C, respectively (calculation and diagram with courtesy from D. Gessmann).

CD spectra in different detergents were recorded for hTom40A. These detergents included β-OG, oPOE and DDM (see also Table 9) and resulted in similar spectra (data not shown). Calculation of secondary structure with CDpro gave related values and concludes the efficient refolding in various detergents. However, it must be mentioned that secondary structure calculations from CD spectra have limitations in accuracy when not coupled with structural data from crystallography or NMR (Khrapunov 2009). This matter will be addressed later in this work (see 4.5).

To assess the thermal stability of hTom40AΔ1-82 and hTom40BΔ1-29 both proteins were denatured by heating from 20 to 95 °C and CD spectra were taken in steps of 10 °C. The high thermal stability of hTom40A and B ($T_m$ hTom40A = 73.1 °C and $T_m$ hTom40B = 74.1 °C; Figure 3.12 C and D) further supported the efficient refolding.

Even though the secondary structure calculation by CDpro might be limited it can reveal changes in secondary structure content during protein heating. An interesting observation was obtained during the analysis of the CD spectra of human Tom40s when recorded just below the melting point of the protein at 70 °C. Both CD spectra of human Tom40 isoforms underwent a remarkable right-shift in combination with decreased ellipticity values, by comparison to CD spectra recorded at 20 and 50 °C Figure 3.12 A, B, dashed lines). Determination of secondary structure content displayed a substantial loss in α-helix content with an increase in β-sheet portion. For hTom40A and B the α-helix content decreased from 20 to 5 % and from 24 to 12 %, respectively. The corresponding relative β-sheet content increased from 36 to 43 % and 32 to 37 %. The relative amount of random coil and turn structure did not change significantly with temperature. Secondary structure development of hTom40A (Figure 3.12 A) and B (Figure 3.12 B) during melting showed that first the α-helical portion is reduced during temperature increase which indicates that the N-terminal helix is denaturing before the barrel-portion follows. Thus, the truncated amino acids of the Tom40 isoforms might not be involved in the overall barrel structure and rather exhibit different functions as discussed below (see 4.4). Again, hTom40B showed little differences upon heating, whereas the overall transition was similar to hTom40A (Figure 3.12 A, B).

### 3.2.2.3 Secondary structure determination by FTIR spectroscopy

To determine complementary data for secondary structure and to support the CD spectroscopy data, refolded hTom40AΔ1-82 and hTom40BΔ1-29 were analysed by Fourier transformation infra-red (FTIR) spectroscopy. The protein samples were needed in high concentrations for an intense signal in the range of wavenumbers from 1700 – 1600 cm$^{-1}$. A high protein concentration of > 5 mg/ml was achieved by purification via a 20 ml Ni-NTA column and subsequent dialysis to remove imidazole. Additional concentration of the protein with spin-columns had to be avoided as even intensive dialysis after concentration could not adjust the detergent concentration in the sample according to the reference buffer. The detergent of choice for these measurements was LDAO as it allowed high protein concentrations for both isoforms. The resulting fractions from the Ni-NTA elution contained 7 mg/ml for hTom40A and 5 mg/ml for hTom40B and were sufficient for intense signals.

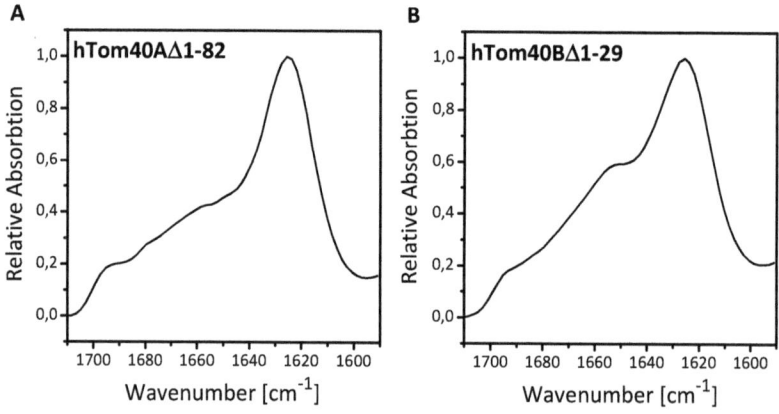

Figure 3.13: FTIR spectra of hTom40AΔ1-82 and hTom40BΔ1-29. The IR-spectra of hTom40A (~ 7 mg/ml) and hTom40B (~ 5 mg/ml) represent an average of three independent spectra with 32 consecutive scans each. Spectral bands assigned to α-helix structure are centred at a wave number of about 1650 cm$^{-1}$, random components at 1645 - 1640 cm$^{-1}$, and β-sheet at 1630 - 1625 cm$^{-1}$. The shoulder at 1695 cm$^{-1}$ indicates antiparallel β-sheet with particularly short turns. For all spectra, the baseline was subtracted. The main peak at 1625 cm$^{-1}$ for hTom40A and hTom40B, respectively, is indicating high β-sheet content.

The amide I bands (1700 – 1600 cm$^{-1}$) of the IR spectra of both isoforms are shown in Figure 3.13. The maximum of both spectra was at 1632 cm$^{-1}$ which is indicative for β-sheet content. The spectra show a clear dominance of β-sheet, which stands in comparison with FTIR measurements of other β-barrel proteins like human VDAC1 (Engelhardt et al. 2007). Calculated from reference data of more than 40 proteins with known 3D structure (Bruker Protein Spectra Library) the amount of β-sheet was determined to be > 55 % (Table 11). At wavenumbers of 1650 cm$^{-1}$ the spectra show no significant local maximum. Comparisons with reference spectra indicate a α-helical content of below 10 %. In agreement with CD spectroscopy data, FTIR spectra showed that the content of β-sheet is dominant in the structure while the α-helical amount is relatively low. The additional maximum at 1656 cm$^{-1}$ is indicating anti-parallel β-sheet and gives additional hint at the organisation of the β-sheets in the protein. This peak is more pronounced in the spectrum of hTom40B. The secondary structure determinations by FTIR support the hypothesis of Tom40 being composed of a β-barrel with a α-helical elongation at the N-terminus.

### 3.2.2.4 Electron microscopy of hTom40

For further structural information about the conformation of human Tom40AΔ1-82 purified protein in different detergent solution was analysed by electron microscopy to evaluate any influence of detergent on structural organisation. The digital electron microscopic images were sorted according to visible particles which would refer to protein channels. Particles showing one- or two-pore structures were visible in protein solution containing 0.1 mg/ml hTom40A in 0.1% LDAO or 0.05 % Brij. In detergents like oPOE or β-OG no or little amounts of particles were visible.

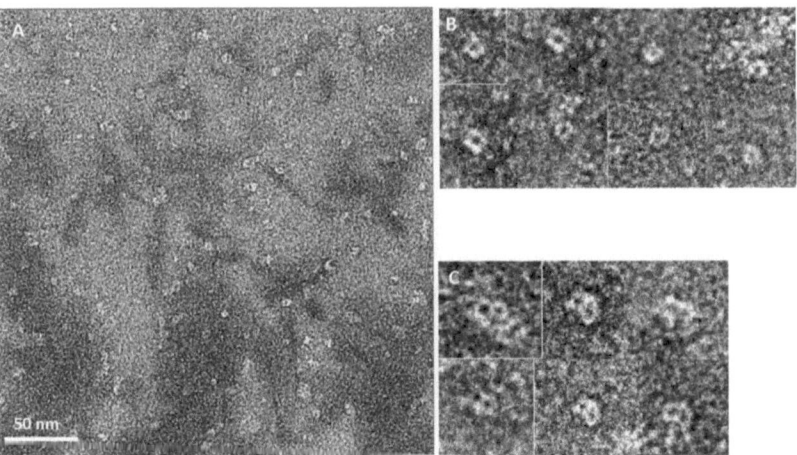

Figure 3.14: EM images of hTom40AΔ1-82 (~ 0.1 mg/ml) in 0.1 % LDAO taken at a magnification of 60.000 x. (A) Survey view of negatively stained hTom40A, (B) images of one-pore particles; (C) images of two-pore particles.

The EM pictures did not reveal a difference in particle size for hTom40A in detergents like LDAO or Brij. The quality of the EM pictures was not appropriate to identify a sufficient amount of single particles for a classification analysis. However, a selection of one-and two-pore structures found in 0.1 % LDAO has been excised (Figure 3.14).

### 3.2.3 Structure modelling of human Tom40

Intensive efforts were made in this project to approach the structure of human Tom40. CD and FTIR showed a high β-barrel content of human Tom40AΔ1-82 and Tom40BΔ1-29. The results of secondary structure analyses with these methods are summarized in Table 11. Bioinformatic analysis of the amino acid sequence with Ali2D (http://toolkit.tuebingen.mpg.de/ali2d) predicted a similar β-strand organization of hVDAC1 and the human Tom40 isoform A and B (Figure 1.3). All 19 β-strands from the solved crystal structure from hVDAC1 could be aligned with the predicted β-strands from hTom40A and B indicating a similar overall structure for both proteins (Bayrhuber et al. 2008; Zeth 2010). Human VDAC1 and Tom40 even share the motif of an N-terminal helix which might support the stability of the β-barrel. The major difference between human VDAC1 and Tom40A and B is the N-terminal elongation in human Tom40 which comprises 80 residues in the isoform A and 30 residues in isoform B. This elongation is not present in hVDAC1 (Figure 1.3).

Table 11: Secondary Structure of hTom40AΔ1-82 and hTom40BΔ1-29 in comparison with hVDAC1

|  | CD | | FTIR | | Psipred[#] | |
| --- | --- | --- | --- | --- | --- | --- |
|  | α-Helix | β-Sheet | α-Helix | β-Sheet | α-Helix | β-Sheet |
| hTom40A | 20 | 36 | nd | 62 | 8 | 52 |
| hTom40B | 24 | 32 | 8 | 57 | 8 | 54 |
| hVDAC1* | 18 | 40 | 7 | 48 | 0 | 63 |

Values in [%],
*(Engelhardt et al. 2007; Malia and Wagner 2007), [#](Jones 1999; McGuffin et al. 2000).

CD and FTIR measurements have been performed to evaluate the folding state of recombinant human Tom40. Data from CD and FTIR measurements for human VDAC1 are highly comparable with the results generated in this work. The secondary structure contents calculated with CDpro for human Tom40A/B and human VDAC1 show the inaccuracy of this algorithm when compared with secondary structure calculated from FTIR data. The α-helical content calculated from CD spectroscopy for hVDAC1 is calculated too high and therefore it can be surely assumed that the α-helical content for hTom40A is predicted too high, as well. The β-sheet contents for all three proteins from FTIR measurements are comparable and resemble the most accurate prediction. Interestingly, the α-helical content of hVDAC1 according to prediction with Psipred indicates no α-helical content but this seems to be an artifact as the crystal structure of hVDAC1 shows the N-terminal helix.

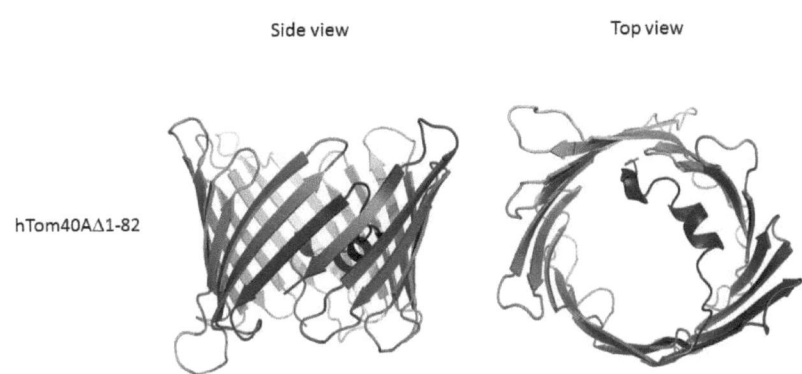

Figure 3.15: Structural model of humanTom40AΔ1-82 (Zeth 2010)

The model was made with Pymol based on the structure of mouse VDAC1 (Ujwal et al. 2008) by Kornelius Zeth and Andrea Guanera (MPI Tübingen) with the N-terminal helix facing inside the pore, A: side view, B: top view. The N-terminal extension has not been modelled as it is predicted to form a random coil.

Based on the structure of human VDAC1, a structural model of hTom40A was made by aligning the two proteins (Bayrhuber et al. 2008; Ujwal et al. 2008). The significant similarity in the secondary structure between hTom40 and hVDAC1 gave reliability to build a structural model of human Tom40 based on the crystal structure of hVDAC1 (Figure 3.15, (Zeth 2010). The β-strands in the model human Tom40 fit perfectly to the structure of VDAC1 besides some gaps which mostly occur in loop regions. The N-terminal elongation is not present in the model as it is predicted to be unstructured. The position of the N-terminal α-helix is shown to be in the barrel lumen. However, this position is not proven and might even be variable.

### 3.2.3.1 Channel activity of hTom40AΔ1-82 and hTom40BΔ1-29

To test whether purified hTom40AΔ1-82 and hTom40BΔ1-29 are active and capable of conducting ion flux, the isolated and refolded proteins were inserted in artificial planar lipid bilayers. The channel-forming activity of the proteins was tested upon application of different voltages and current traces were recorded. The traces were measured with a single channel as well as multiple channels inserted in the artificial lipid layer. For single channel characterization a planar membrane was reconstituted and a baseline was measured without any voltage applied. Protein was added to the cis-side of the chamber while a voltage of 30 mV was applied and current traces were recorded. Single channel insertions have been observed (Figure 3.16 A) and were analysed for mean single-channel conductance. All single channel events for human Tom40AΔ1-82 showed a peak conductance at ~ 1.2 nS (Figure 3.16 B). The channel insertion events of hTom40A were rare and hard to obtain. As channel insertions were even less frequent for hTom40BΔ1-29 no definite single channel conductivity could be determined for human Tom40B.

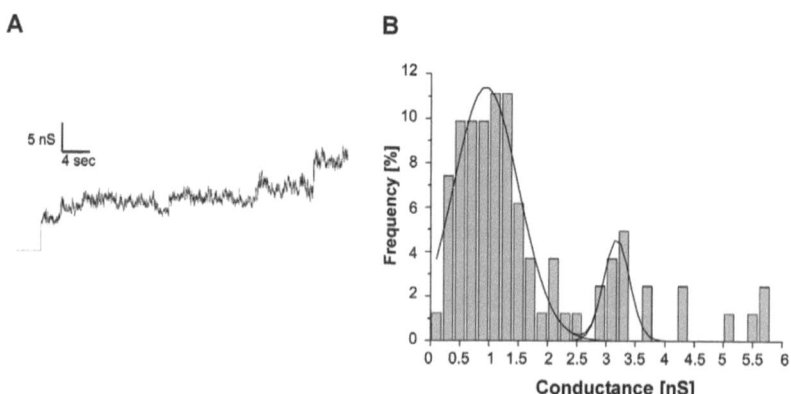

Figure 3.16: (A) Purified hTom40AΔ1-82 was added to both sides of a planar lipid membrane formed by DiphPC/n-decane/butanol and single channel conductances were measured in the presence of a membrane potential of 30 mV. (B) Histogram of channel conductivity of hTom40AΔ1-82 measured at voltages between ±10 and ±50 mV. A total of n = 82 conductance increments were analysed. The aqueous phase contained 1 M KCl, 50 mM HEPES, pH 7.2. Most channel insertions showed a conductance of 1.2 nS.

To test whether the inserted channels show voltage-dependent gating, several hTom40AΔ1-82 and hTom40BΔ1-29 channels were allowed to reconstitute into planar lipid membranes and steps of different voltages starting from ±10 mV and ending at ±160 mV were applied. Baseline currents at 0 mV were always measured in between to determine ΔI for the respective voltage. For human Tom40AΔ1-82 and Tom40BΔ1-29 the recorded currents showed no voltage dependent gating at voltages below 120 mV. The total conductivity of multiple channels, which was measured at different voltages, remained the same. For hTom40A the measurements showed a slight decrease of conductivity at voltages higher than 120 mV (Figure 3.17, A). Also, for hTom40B no voltage dependent gating was observed even at voltages above 120 mV (Figure 3.17, B). Predominantly, a voltage independent behaviour was observed for both isoforms which stand in contrast to the voltage dependence of hVDAC1 which shows a channel closure at voltages above 50 mV (Engelhardt et al. 2007).

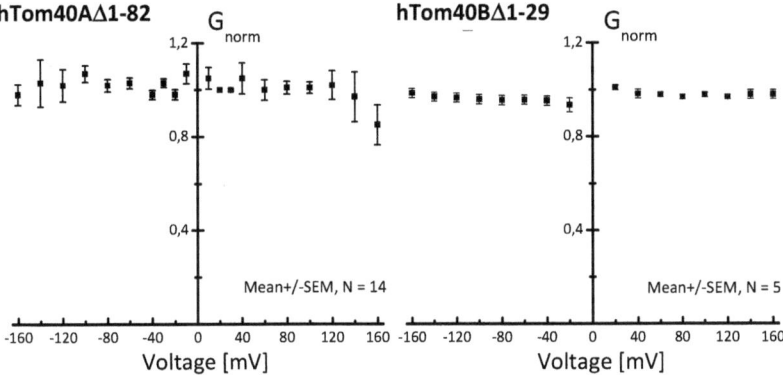

Figure 3.17: Voltage dependence of human Tom40

hTom40AΔ1-82 and hTom40BΔ1-29 were reconstituted into black lipid membranes and conductivity was determined between -160 and +160 mV. $G_{norm}$ represents the normalized channel conductance and was calculated according to G(V)/G(10 mV). Human Tom40AΔ1-82 shows a minimal voltage conductance at voltages above 100 mV, hTom40BΔ1-29 shows no voltage dependence even at high voltages above 120 mV.

## 3.3 Stability of β-barrel membrane proteins

After having established the purification and folding of hTom40AΔ1-82 and hTom40BΔ1-29 for structural studies it has been decided to prepare also a stabilized Tom40 protein that might be better suited for structural studies based on x-ray crystallography. Crystallization of membrane protein has been very successful with proteins isolated from thermophilic organisms that had high thermal stability (Aartsma and Matysik 2008).

In a collaboration project with Dennis Gessmann (Dept Biophysics, Stuttgart University), Hammad Naveed and Jie Liang (Dept. of Bioengineering, University of Illinois, Chicago, USA) I was looking for a stabilized mutant of hTom40AΔ1-82. The stability of β-barrel membrane proteins is determined by the balance between favourable hydrogen-bonding, van der Waals and hydrophobic interactions as well as unfavourable conformational entropy. To identify unstable regions in the transmembrane domain of hTom40AΔ1-82 the group of Jie Liang has estimated the energetic contribution of all β-strands of human Tom40A. This approach has recently been applied to model the conformational stability of 25 nonhomologous β-barrel membrane proteins with known structure (Naveed et al. 2009).

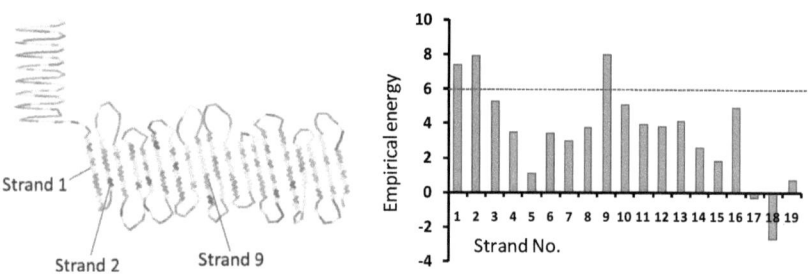

Figure 3.18: Left: A model of the protein topology of hTom40A was generated using Psipred predictions (Jones 1999; McGuffin et al. 2000) and TMRPres2D (Spyropoulos et al. 2004). Right: Empirical energy of β-strands 1 - 19 of wild type protein: β-strands 1, 2 and 9 are predicted as the weakly stable strands. The dotted line represents threshold above which energy level indicates weakly stable strands (Figure with courtesy from J.Liang).

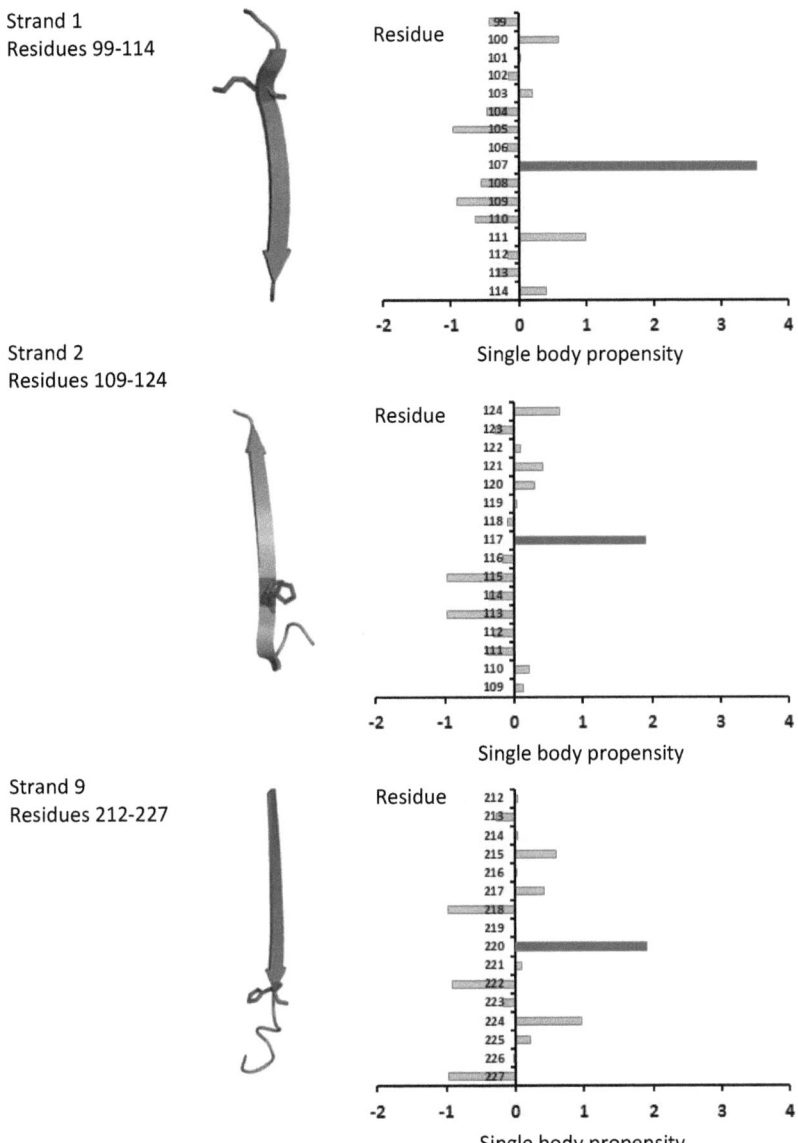

Figure 3.19: Weakly stable β-strands of hTom40A and unstable amino acids are coloured in dark grey. Single body propensities of amino acid residues of β-strands 1, 2 and 9 show amino acids K107, H117, and H220 to have the largest values indicating a destabilizing effect on the regarding strand (Figure with courtesy from J. Liang).

Based on these energy calculations, the energy contribution of the overall β-structure of hTom40AΔ1-82 identified three weak β-strands, strands 1, 2 and 9, which have significantly higher energies and are thus less stable then the remaining strands of the protein (Figure 3.18). Further examinations of the energy contribution analyzing only the amino acids in the three strands 1, 2 and 9 identified one amino acid in each strand being responsible for the high energy level. The residues K107 in strand 1, H117 in strand 2, and H220 in strand 9 were predicted be most responsible for the instability of these strands (Figure 3.19, Figure 3.20).

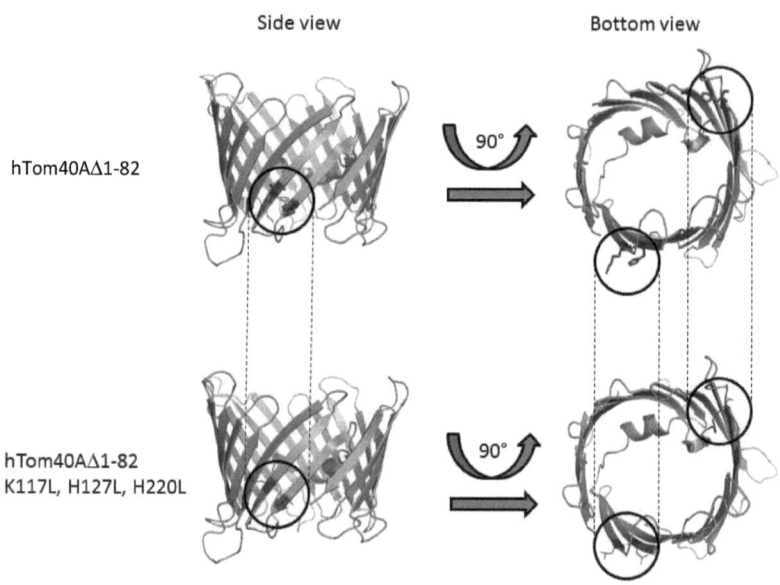

Figure 3.20: Structural overview of the three mutated amino acids in a model of human Tom40AΔ1-82 wild type and mutant. The three destabilizing amino acids are marked in dark grey in the wild type protein in the upper model (top view and bottom view). The three mutated amino acids are marked in dark grey in the mutant protein in the lower model (top view and bottom view).

### 3.3.1 Cloning, expression and refolding of mutated hTom40AΔ1-82

After the theoretical energy contribution has identified three destabilizing amino acids in hTom40AΔ1-82 the attempt to mutagenize these amino acids was made. The three amino acids K107 in strand 1, H117 in strand 2, and H220 in strand 9 were replaced by leucines with site-directed mutagenesis. As the first two amino acids in strand 1 and 2 were close together in the gene coding for hTom40AΔ1-82 one primer for both mutations was sufficient. Nevertheless, primers for the single mutations have been used as well and single mutants are still available for further examination. The plasmid containing the double mutant K107L and H117L was mutagenized with the primer for the replacement of H220L. Mutagenesis was successful by first attempt and sequence analysis of the resulting primers revealed the correct replacement of all three amino acids. All following experiments were carried out with the triple mutant K107L, H117L, K220L. Transformation was done with *E. coli* C41 cells as these showed better growth behavior than BL21.

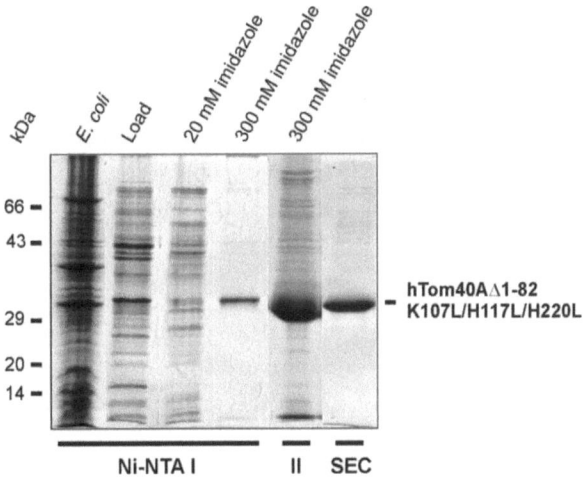

Figure 3.21: SDS-PAGE showing mutant (K107L, H117L and H220L) human Tom40A, lacking the N-terminal residues 1-82. The protein was expressed in *E. coli*, purified from inclusion bodies under denaturing conditions (6 M guanidine chloride) and and loaded onto a Ni-NTA affinity column. Human Tom40A was eluted under denaturing conditions with 300 M imidazole (Ni-NTA I) and refolded by rapid dilution into 0.5 % LDAO. Refolded protein was subjected to Ni-NTA chromatography (Ni-NTA II) and eventually passed over a Superose12 size-exclusion column (SEC). Aliquots of the resulting column fractions were analyzed by SDS-polyacrylamide gel electrophoresis and Coomassie blue staining. Lane 1: purified inclusion bodies from *E. coli* cells expressing hTom40A K107L, H117L and H220L, lanes 2 to 5: Ni-NTA column fractions and imidazole eluates; lane 6: peak fraction of SEC.

The triple mutant K107L/H117L/K220L, also termed hTom40AΔ1-82$^{3mut}$, was expressed in *E. coli* C41 and cells were harvested and lysed according to human Tom40A wild type (see 3.2.1). The first purification under denaturating conditions removed most of the impurities. A refolding screen was performed for hTom40AΔ1-82$^{3mut}$ which revealed similar results compared to human Tom40A wild type (Table 12). The detergents LDAO and Brij35 gave best results for protein stability and showed little to no aggregation after 170 h. To ensure comparability between the wild type Tom40 and the mutant all experiments were performed under the same conditions (see Table 9).

Table 12: Yield of soluble hTom40AΔ1-82$^{3mut}$ after refolding

|  | hTom40AΔ1-82$^{3mut}$ 1 h | | | | hTom40AΔ1-82$^{3mut}$ 24 h | | | | hTom40AΔ1-82$^{3mut}$ 170 h | | | |
|---|---|---|---|---|---|---|---|---|---|---|---|---|
| Detergent | pH 6.0 | pH 7.0 | pH 8.0 | pH 10.0 | pH 6.0 | pH 7.0 | pH 8.0 | pH 10.0 | pH 6.0 | pH 7.0 | pH 8.0 | pH 10.0 |
| 0.05 % DDM | 100 | 100 | 100 | 15 | 100 | 80 | 55 | 65 | 95 | 45 | 10 | 50 |
| 0.1 % LDAO | 100 | 100 | 100 | 100 | 100 | 100 | 95 | 100 | 100 | 100 | 95 | 100 |
| 0.5 % OPOE | 70 | 15 | 35 | 100 | 15 | 10 | 25 | 10 | 10 | 10 | ≤5 | 10 |
| 0.05 % Brij35 | 100 | 100 | 100 | 60 | 75 | 85 | 100 | 100 | 65 | 75 | 70 | 95 |
| 3 % CHAPS | 60 | 35 | 35 | 100 | 25 | 20 | ≤5 | 100 | 15 | ≤5 | ≤5 | 15 |
| 1 % β-OG | 30 | 15 | ≤5 | 70 | 10 | ≤5 | ≤5 | ≤5 | 10 | ≤5 | ≤5 | ≤5 |

The detergent concentrations are given in w/v, the protein concentration is given in % of total.

### 3.3.2 Secondary structure determination of hTom40AΔ1-82$^{3mut}$

To examine correct protein folding the secondary structure content of mutant human Tom40A was analysed by CD spectroscopy and FTIR (Figure 3.22). The CD spectrum shows a clear dominance of β-sheet content. The minimum of the spectra consistently clustered around 216 nm with a crossover of the baseline at 203 nm. At wavelengths > 248 nm, the CD spectrum approached ellipticity values close to zero. Secondary structure prediction of the mutant human Tom40A with the amino acid exchange in K107L/H117L/K220L showed similar results as for the wild type. CD spectra for both proteins look the same. The secondary structure content calculated with CDpro (Sreerama and Woody 2000; 2003; 2004) revealed a predominance of β-sheet secondary structure (> 35 %) for hTom40AΔ1-82$^{3mut}$ with low α-helical content (< 20 %).

Secondary structure content calculated from reference pattern for FTIR revealed a β-sheet content of 54 % and a α-helical content of 1.5 %. These results were in good agreement with the data obtained for the wild type hTom40AΔ1-82.

Both secondary structure measurements revealed a readily folded protein which was suitable to be tested for increased thermal and chemical stability compared to the wild type protein.

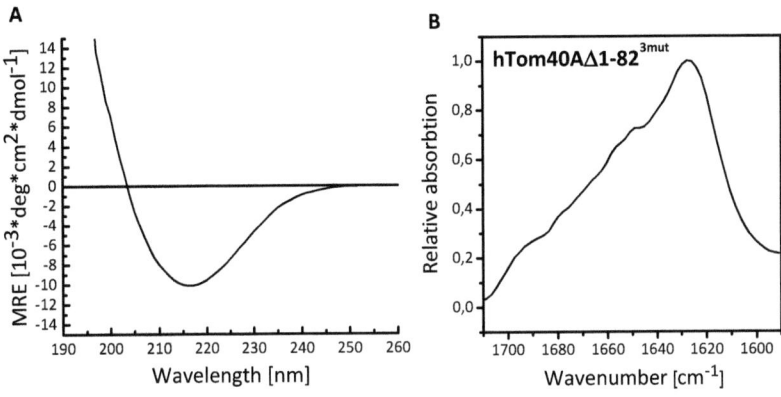

Figure 3.22: CD spectrum and FTIR spectrum of hTom40AΔ1-82 K107L/H117L/K220L

A: hTom40A Δ1-82 K107L/H117L/K220L (~0.2 mg/ml) was solubilized in 0.5 % LDAO, 20 mM Tris, pH 8.0. For each CD spectrum, 5 scans were accumulated at 20 °C (solid line), 50 °C (dashed line) and 70 °C (dashed line) and background corrected. Noisy data below 200 nm have been removed. Mean residue ellipticity (MRE) was calculated based on the molar protein concentration and the number of amino acid residues of the according protein. The spectrum indicates a high ratio of β-sheet with a minimum at 216 nm and a crossover of the baseline at 203 nm.

B: The IR-spectra of hTom40A Δ1-82 K107L/H117L/K220L (~ 3 mg/ml) represents an average of four independent spectra with 32 consecutive scans each. Spectral bands assigned to α-helix structure are centred at a wave number of about 1650 cm$^{-1}$, random components at 1645 - 1640 cm$^{-1}$, and β-sheet at 1630 - 1625 cm$^{-1}$. The main peak at 1625 cm$^{-1}$ for hTom40A K107L/H117L/K220L is indicating high β-sheet content. The shoulder at 1695 cm$^{-1}$ indicates antiparallel β-sheet with particularly short turns. For all spectra, the baseline was subtracted.

### 3.3.3 Heat stability of hTom40AΔ1-82³ᵐᵘᵗ

The CD spectrum of mutant Tom40A revealed a readily folded protein suitable for stability tests. To assess the thermal stability of mutant hTom40A the protein was heated according to wild type hTom40AΔ1-82 wild type (see 3.2.2.2) and revealed a melting temperature of 84 °C. This melting temperature was significantly higher than the melting temperature of wild type hTom40AΔ1-82 which had a melting point at 72 °C (see Figure 3.12). This result strongly indicates an increased thermal stability of hTom40AΔ1-82$^{3mut}$ caused by the mutations of the three destabilizing amino acids K107L, H117L, K220L in the strands 1, 2 and 9, respectively.

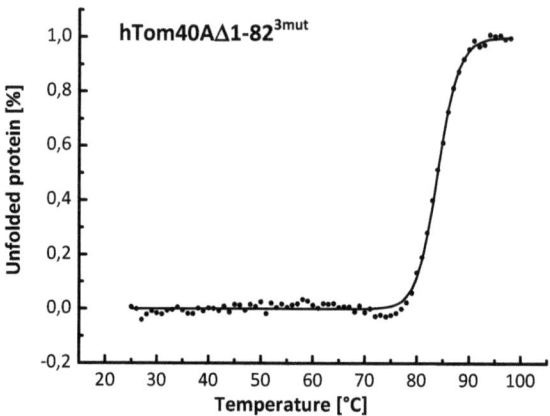

Figure 3.23: Thermal stability of hTom40AΔ1-82 K107L/H117L/K220L

Melting curve of hTom40AΔ1-82 K107L/H117L/K220L determined from corrected CD signals measured at a wavelength of 216 nm. The melting temperature of hTom40AΔ1-82 K107L/H117L/K220L was at 84 ± 1.1 °C and therefore 12 °C higher than the melting temperature of hTom40A wild type (see Figure 3.12 B).

### 3.3.4 Stability of hTom40AΔ1-82³ᵐᵘᵗ in chaotropic reagents

The improved stability of the hTom40AΔ1-82 K107L/H117L/K220L has been successfully shown by melting the protein under CD spectroscopy monitoring. To further confirm this result a second approach was made to test the stability of the mutated protein in chaotropic reagents like guanidine-hydrochloride. This has been assessed via tryptohan-fluorescence which changes according to the folding state of the protein. Dilution of the protein in buffers containing various concentrations of guanidine-hydrochloride ranging from 0.3 M to 7 M and subsequent fluorescence analysis revealed a difference in the fluorescence pattern between wild type and mutant hTom40A.

Human Tom40AΔ1-82$^{3mut}$ showed a shift in midpoint of fluorescence to a higher guanidine-hydrochloride concentration towards the wild type. While the wild type hTom40A already unfolded at a concentration of 3.3 M guanidine-hydrochloride the mutant protein showed a midpoint shift at 5.3 M. The critical unfolding concentration of guanidine-hydrochloride for wild type and mutant protein differs in more than 2 M guanidine-hydrochloride and remarkably showed the increased stability of the mutant protein in chaotropic reagents.

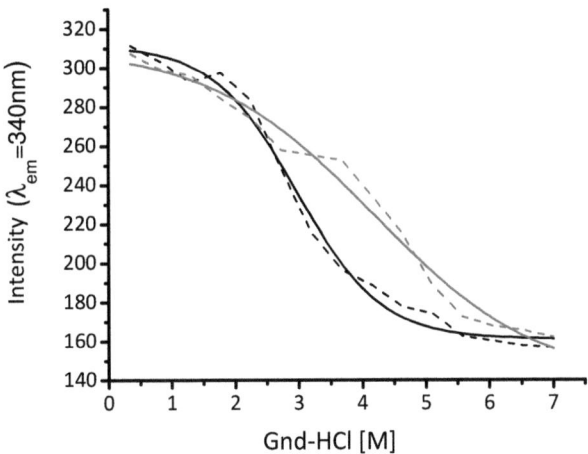

Figure 3.24: Chemical stability of mutant hTom40AΔ1-82 towards wild type hTom40AΔ1-82. Both proteins were dialyzed in 20 mM Tris-HCl pH 8, 1 mM β-ME, 1 % LDAO, 7 M guanidine-hydrochloride and then diluted 1:20 in buffers containing 0-7 M guanidine-hydrochloride (Gnd-HCl). The folding state of the proteins was assessed by determining the fluorescence of tryptophan at 340 nm. Mutant hTom40A (gray) shows an increased stability towards guanidine-hydrochloride as it unfolds at higher guanidine-hydrochloride concentrations than wild type hTom40A (black). A Boltzman fit (solid line) was applied to the data points (dashed line).

### 3.3.5 Crosslinking of wild type and mutant hTom40AΔ1-82

The exchange of amino acids in the weakly stable β-strands was proposed not only stabilize the protein but also to prevent oligomerization. Crosslinking experiments with glutaraldehyde were performed with wild type and mutant hTom40AΔ1-82 to compare their oligomerization state. With this approach subunit composition can be deduced as crosslinkers only bind surface amino residues in relatively close proximity in the native state. Protein interactions are often too weak or transient to be easily detected, but by crosslinking, the interactions can be captured and analyzed. Glutaraldehyde can form stable inter- and intra-subunit covalent bonds where maintenance of structural rigidity of protein is important. The length of the glutaraldehyde monomer is about 7.5 Å, while the bridge between two lysines is about 3 Å in length (Salem et al. 2010).

The analysis of crosslinked wild type and mutant hTom40AΔ1-82 by SDS-PAGE and Western blot revealed a remarkable difference of the oligomerization state between the wild type Tom40 and its mutant form. Human Tom40A wild type forms tight dimers (see lane 1 (-) in Figure 3.25). Incubation with the crosslinker emphasized this band and with further incubation time a trimer band could be identified as well. A different result was revealed for the triple mutant where no bands according to dimers or trimers could be detected. However it should be noted, at longer incubation time the monomeric band fades and more protein was found in the loading pocket of the SDS-gel. This was likely due to the fact that even the mutant protein gets crosslinked even though not localized in close connection.

Remarkably as well, was the fact that mutant hTom40AΔ1-82 was running at a little lower size than the wild type protein hTom40AΔ1-82 in the SDS-gel. Both proteins have almost exactly the same size and both proteins should run equally in the SDS-gel. An explanation for this observation could be that the mutant protein is more stable compared to the wild type so it is not unfolded completely by the SDS in the gel and therefore maintains a more compact structure than the wild type protein which allows faster migration in the gel matrix.

It has to be mentioned that in the mutant protein one lysine was exchanged to a leucine. One could argue that this exchange is responsible for the different crosslinking behavior. Nevertheless, the wild type hTom40A comprises 15 lysines of which one has been mutated to leucine. Although depending largely on the position of the lysine this single exchange would not result in such a remarkable difference in the crosslinking result.

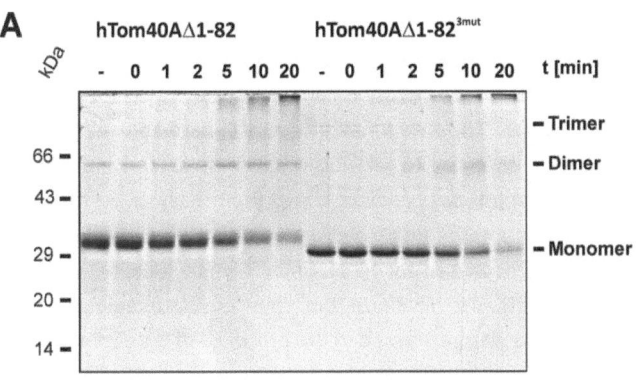

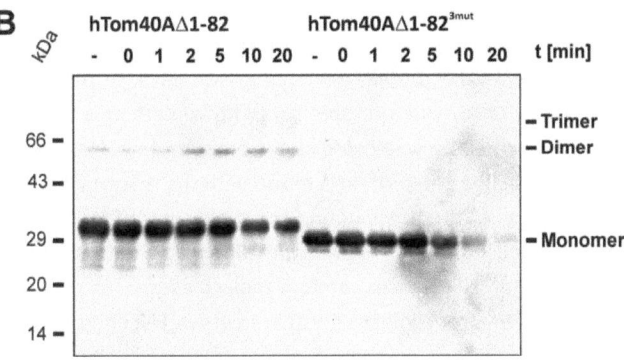

Figure 3.25: Crosslinking of wild type and mutant hTom40AΔ1-82 with glutaraldehyde

The crosslinking reactions were performed in 20 mM Na-Phosphate buffer pH 8, and incubated with 125 µM glutaraldehyde at 37 °C for 0 - 45 min. The concentrations for wild type and mutant protein were 40 µM. Aliquots were removed before (-) and after the addition of the crosslinking reagent at the time indicated. Crosslinking reactions were stopped by addition of 50 mM Tris-HCl pH 8. Proteins were visualized by (A) coomassie staining and (B) Western blotting with antibodies for mammalian Tom40 (see Table 7).

### 3.3.6 3D crystallization of wild type and mutant hTom40AΔ1-82

The structural determination of membrane proteins is challenging due to their hydrophobic nature and their non-native solubilization in detergent solution. The difficulties in crystallizing membrane proteins with rather large hydrophobic domains is hindered as only few protein-protein contact sites are available for crystal packing (Caffrey 2003). However, the structures of several β-barrel proteins were determined by 3D-crystallization. In order to provide detailed structural information regarding human Tom40A attempts to crystallize the protein have been made in this work. Both, the wild type human Tom40Δ1-82, and the mutant hTom40Δ1-82, in which three residues have been replaced to confer higher stability, were subjected to crystallization attempts. To maximize the chances of obtaining highly diffracting crystals, crystallization was performed in the presence of detergent micelles (3.3.6.1), in detergent/lipid bicelles (3.3.6.2), and in a lipidic cubic phase (3.3.6.3).

#### 3.3.6.1 Crystallization in detergents

All crystallization experiments were performed using the vapour-diffusion method (McPherson, 1978), with the protein in either a "sitting" and "hanging" drop in the vapour-diffusion. The underlying principle behind this method is that protein super-saturation can be attained by water diffusion from a protein drop to a precipitating agent reservoir. In practice, the protein is mixed with the reservoir solution at a given ratio, and equilibration of the precipitating agent concentration in the drop and the reservoir is achieved through water diffusion. As the precipitating agent concentration in the drop increases, protein molecules may form a scaffold with a certain order – a nuclei – from which a crystal can grow. The compelling advantage of the vapour-diffusion method is that it allows screening through thousands of crystallisation conditions. This process can furthermore be robotized. The most commonly used precipitants for the crystallization of proteins are polyethylene glycols (PEGs) and salts (Hunte 2003).

Crystallization trials with hTom40AΔ1-82 in detergent solution were set up at the Max Planck Institute (MPI), Tübingen, in the Department of Protein Evolution by robot-assisted pipetting in 96-well plates. The protein concentration used in these trials ranged between 3 and 8 mg/ml and conditions of over 1300 different precipitant solutions have been tested (see 2.5.8.1). The detergents used in these trials were 0.5 % LDAO or 1 % oPOE. Crystal growth was achieved with hTom40AΔ1-82 (7 mg/ml) in 20 mM Tris-HCl pH8, 1 mM β-ME and 1 % LDAO in several precipitant conditions, predominantly in a pH range of 7 to 9 (Figure 3.26). Additives supporting crystal growth were jeffamine or high molecular mass PEGs. Crystals appeared in less than 1 week and were left growing for 3 weeks. To test the diffraction of the grown crystals they were harvested in a nylon loop and frozen in liquid nitrogen before exposure to x-ray beams at the PX10 beamline in the Swiss Light Source synchrotron. The diffraction of the tested crystals and their buffer conditions is listed in Table 13. Best crystal diffraction of 11.5 Å was achieved from crystals grown in 100 mM HEPES pH 7 and 30 % jeffamine (Table 13, Figure 3.26, A).

Table 13: Crystals of hTom40AΔ1-82 from screens at the MPI

| Figure 3.26 | Screen | Company | Salt | Buffer | pH | Additive |
|---|---|---|---|---|---|---|
| A | Screen Index | Hampton | - | 0.1 M HEPES | 7 | 30 %(v/v) Jeffamine |
| B | Screen Index | Hampton | 20 mM trimethylamineoxide | 0.1 M Tris | 8.5 | 20 %(v/v) PEG 2000 |
| C | BS Wizard III | Emerald BioSystems | - | 0.1 M MES | 6.5 | 20 %(v/v) PEG 1500 |
| D | PACT | Quiagen | - | 0.1 M SPG | 7 | 25 %(v/v) PEG 1500 |
| E | PACT | Quiagen | - | 0.1 M SPG | 8 | 25 %(v/v) PEG 1500 |
| F | PACT | Quiagen | - | 0.1 M SPG | 9 | 25 %(v/v) PEG 1500 |

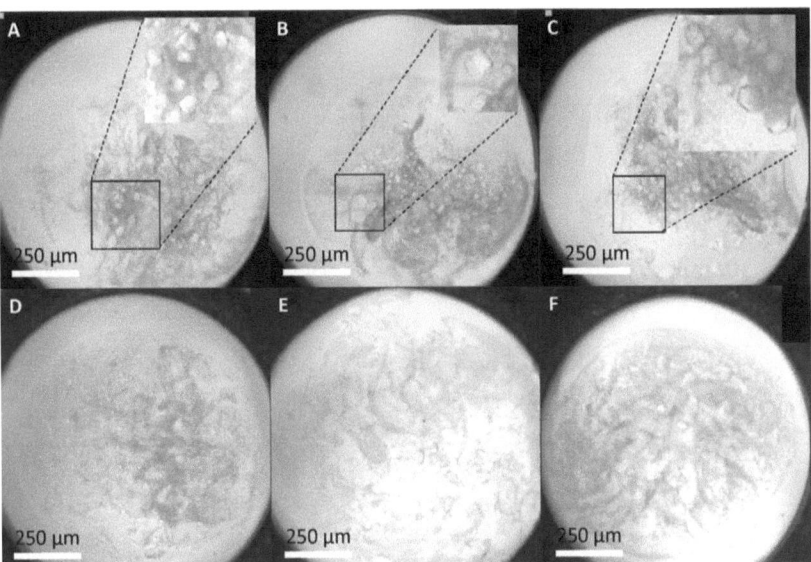

Figure 3.26: Crystals of hTom40A Δ1-82 (7 mg/ml) in 1 %LDAO, Crystals were grown in buffer solutions containing precipitants listed in Table 13.

While the diffraction patterns obtained from these crystals indeed confirmed their proteinaceous nature, they were insufficient to determine any space groups. In an attempt to improve the diffraction quality, an additive screen (Hampton Additive Screen HR2-138) was performed which failed in improving crystal quality. Following attempts to repeat the screens did not show any further crystal growth.

For a broader attempt to obtain crystals of hTom40AΔ1-82 collaboration was started with Dr. Jacques-Phillipe Colletier from the Structural Biology Institute in Grenoble. There, the methods were enlarged not only to crystallization trials with detergent in precipitant buffers but as well to crystallization in bicelles and lipidic cubic phases.

Table 14: Crystals obtained from hTom40AΔ1-82 in detergent solution at the ESRF

| Figure 3.27 | Screen | Company | Salt | Buffer | pH | Additive |
|---|---|---|---|---|---|---|
| A | The Classics | Quiagen | 0.2 M MgCl$_2$ | 0.1 M Tris | 8.5 | 3.4 M 1.6-Hexanediol |
| B | The Classics | Quiagen | 0.2 M tri-Na citrate | 0.1 M HEPES-Na | 7.5 | 20 %(v/v) Isopropanol |
| - | The Classics | Quiagen | 0.2 M MgCl$_2$ | 0.1 M HEPES-Na | 7.5 | 30 %(v/v) Isopropanol |
| - | The Classics | Quiagen | 0.2 M NH$_4$ acetate | 0.1 M Tris | 8.5 | 30 %(v/v) Isopropanol |
| C | The Classics | Quiagen | 0.2 M Mg acetate | 0.1 M Na cacodylat | 6.5 | 30 %(v/v) MPD |
| D | The Classics | Quiagen | 0.2 M NH$_4$ PO$_4$ | 0.1 M Tris | 8.5 | 50 %(v/v) MPD |
| E | Natrix | Hampton | 5 mM MgSO$_4$. | 0.05 M Tris | 8.5 | 35 %w/v 1.6-hexanediol |
| F | MembFac | Hampton | 0.1 M MgCl$_2$ 6*H$_2$O | 0.1 M HEPES-Na | 7.5 | 18 %v/v PEG 400 |
| - | MembFac | Hampton | 0.1 M MgCl$_2$ 6*H$_2$O | 0.1 M Na acetate trihydrate | 4.6 | 18 %v/v PEG 400 |
| - | Natrix | Hampton | 0.1 M KCl | 0.05 M Tris | 8.5 | 30 %v/v PEG 400 0.01 M MgCl$_2$ 6*H$_2$O |
| - | Mme 5000 | HTX-Lab | | 0.1 M Citric Acid | 4 | 5% PEG MME 5000 |
| G | Mme 5000 | HTX-Lab | | 0.1 M Citric Acid | 5 | 5% PEG MME 5000 |
| H | MPD | Hampton | | 0.1 M bicine | 9 | 20 %v/v 2-Methyl-2.4-pentanediol |
| - | Screen Index | Hampton | | 0.1 M Na acetate trihydrate | 4.5 | 25 %w/v PEG 3350 |
| I | Screen Index | Hampton | 0.2 M NH$_4$ acetate | 0.1 M bis-tris | 6.5 | 45 %v/v 2-methyl-2.4-pentanediol |

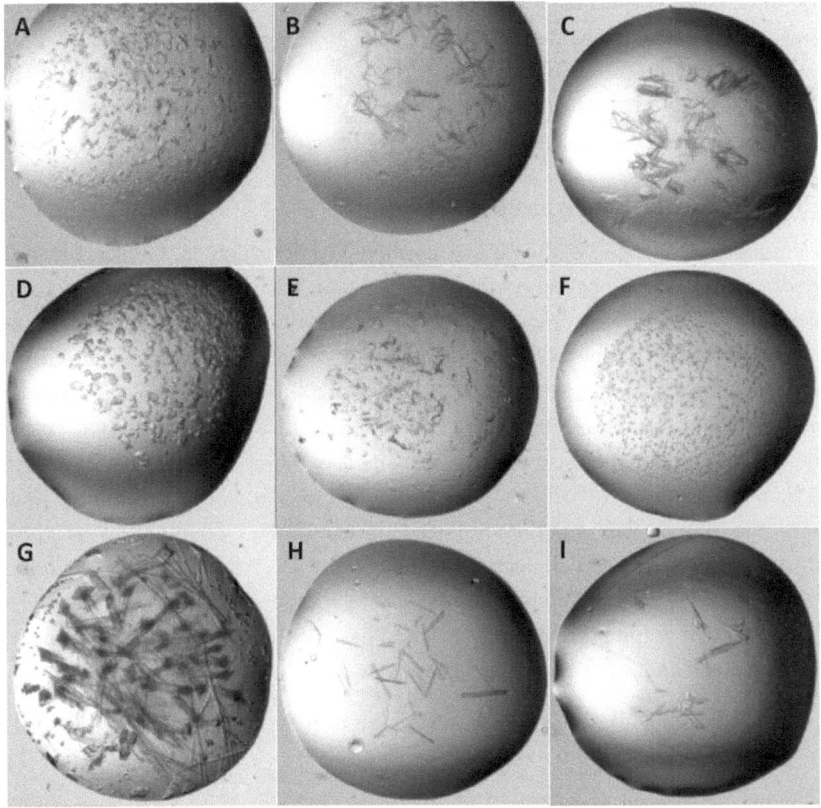

Figure 3.27: Crystals of hTom40A Δ1-82 (10 mg/ml) in 0.1 %LDAO, Crystals were grown in buffer solutions containing additives listed in Table 14. (Drop size 100 nl, ∅ ~300 μm)

The concentration of hTom40AΔ1-82 used in these trials was above 10 mg/ml with a detergent concentration of 0.1 % LDAO in 20 mM Tris-HCl pH 8, 1 mM β-ME and various precipitants have been tested (see 2.5.8.2.). Crystals grown from detergent solutions were obtained in various conditions (see Table 14) and picked for diffraction analysis. To avoid ice formation in the crystals, which would lead to a disruption of crystalline order, mother liquor solutions containing 18% glycerol were prepared for each crystal type beforehand, in which crystals were transferred before their flashcooling to 100 K for data collection. Sufficiently large crystals were mounted in nylon loop and flash-cooled to 100 K directly in a nitrogen gas stream. Crystals, smaller than 5 μm in radius, were mounted in batches at the same time in a kapton-grid loop, before flashcooling them directly in the nitrogen gas stream. The crystals qualities were assessed by their diffraction power following exposure of the crystal to an x-ray beam at the ESRF, on either ID23EH2 microfocus or BM30A beamline. Crystals diffracted best at a resolution of 11.9 Å and were therefore not suited for further structural characterization.

### 3.3.6.2 Crystallization in bicelles

A frequent challenge in membrane protein crystallization is finding the right conditions for solubilizing the protein while keeping its structure and stability intact. Membrane proteins are most stable in their native bilayer environment but mimicking this environment *in vitro* is a tough task. This matter can be solved with the crystallization in bicelles (Faham and Bowie 2002) which are planar stacks of phospholipid membranes. By mixing hTom40AΔ1-82 and DMPC/CHAPSO bicelle solution, the protein may insert into lamellar discs of the lipids, in which it is expected to be more stable and therefore more prone to crystallization (Faham et al. 2005). Once the protein-lipid solution is mixed, it can further be pipetted by robot-assistance. This enables a high-throughput in the screening of crystallization conditions. The only difference to the crystallization of membrane proteins in detergents is that the protein/bicelle solution has to be kept at 4 °C before its mixing with the precipitant solution to avoid phase separation.

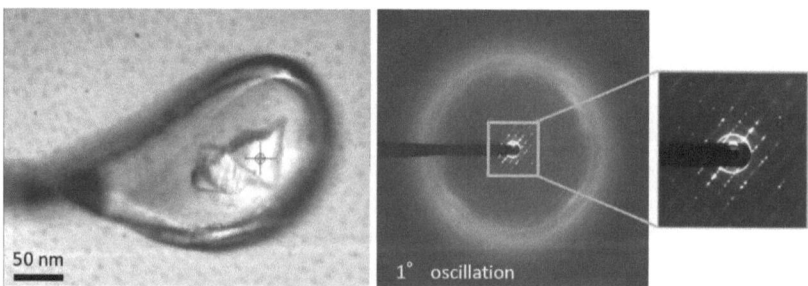

Figure 3.28: Left: Crystal of of hTom40AΔ1-82 (7 mg/ml) grown in 100 mM MES, pH 6.5, 35 % MPD in DMPC/CHAPSO bicelles picked with a nylon loop. Right: Diffraction pattern of a hTom40AΔ1-82 crystal diffracting up to 8 Å.

Crystals of hTom40AΔ1-82 grown in bicelles were harvested in a loop and analyzed at various ESRF beamlines (Figure 3.29). Crystals grown from protein in bicelles showed a diffraction pattern with resolutions between 11 and 60 Å. We conjecture that these patterns are indicative of stacks of lipid bilayers containing hTom40AΔ1-82 protein arranged one over the other (Figure 3.29).

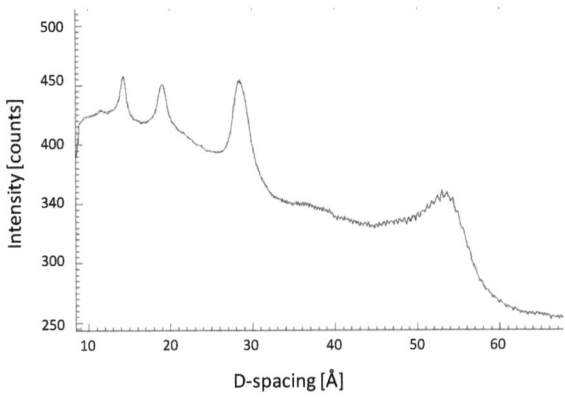

Figure 3.29: Diffraction pattern of hTom40AΔ1-82. The diffraction goes up to 14 Å. The spacing between the rings is ~60 Å (to evaluate it from the D-spacing graph, d(Ang)=10/((10/r1)-(10/r2)) ).This corresponds to membrane spacing and reveals a type 1 membrane protein crystal in which crystal packing interactions between non-hydrophobic domains are random.

### 3.3.6.3 Crystallization in lipidic cubic phase

For crystallization of wild type and mutant hTom40AΔ1-82 in lipidic cubic phases the protein solutions were mixed with lipids in the solid phase. The cubic phase is obtained either following multiples rounds of centrifugation or multiples passes through an emulsifier. The percentage of lipids in the cubic phase can be varied from 50 to 70 %, so 60 % monoolein was used in these trials. The "gel"-like cubic monoolein phase was soaked with precipitant solutions (see 2.5.8.2), which allowed a screening through various conditions (Nollert 2002). This process was automated with robotic systems available at the Structural Biology Institute in Grenoble. As the cubic phase will turn into a "sponge phase" (Caffrey and Cherezov 2009) in the presence of some crystallization precipitants, this technique has been renamed *"in meso"*.

The lipids in cubic phase represent two advantages in regards to other lipids. First, they have no head group that could interact with the protein surface thereby complicating crystallization and second, they form bilayer of comparable thickness to biological membranes. The lipidic cubic phase is a structured, transparent and complex array that is pervaded by two, non-communicating aqueous channel systems. Such matrices permit an easy lateral diffusion of membrane proteins, which may result in the nucleation of crystals. They also provide a good support for the growth of crystals and are more tolerant to impurities than the two before-mentioned crystallization methods. The crystals of hTom40A wild type and mutant grown in screens with 60% monolein cubic phases are shown in Figure 3.30 and Figure 3.31, respectively, in buffer conditions listed in Table 15. The stability mutant hTom40AΔ1-82 K107L/H117L/H220L did show advanced crystal growth compared to wild type hTom40AΔ1-82.

Table 15: Crystals of wild type and mutant hTom40AΔ1-82 in lipidic cubic phases

| Figure 3.30, Figure 3.31 | Screen | Company | Salt | Buffer | pH | Additive |
|---|---|---|---|---|---|---|
| A | MemGold | Molecular Dimensions | 0.1 M Na-Chloride 0.1 M Na-Phosphate | 0.1 M HEPES | 7 | 33 %(v/v) PEG 400 |
| B | MemGold | Molecular Dimensions | 0.05 M Na-Sulfate 0.05 M Li-Sulfate | 0.05 M Tris | 8.5 | 35 %(v/v) PEG 400 |
| C | MemPlus | Molecular Dimensions | 0.5 M Na-Acetate | 0.05 M Tris | 8 | 0.1 M Imidazole 25 % MPD |
| D | MemStart + MemSys | Molecular Dimensions | - | 0.1 M Na-Citrate | 5.5 | 1.5 M Na-Phosphate |
| E | MemStart + MemSys | Molecular Dimensions | 0.2 M Ca-Chloride | - | | 30 %(v/v) Isopropanol |

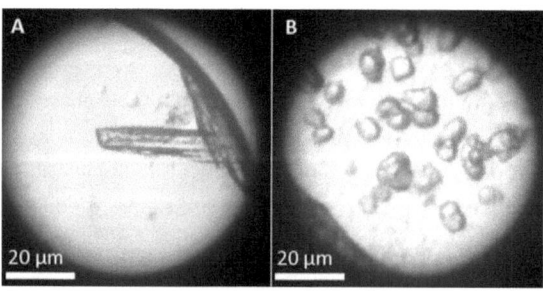

Figure 3.30: Crystals of hTom40A Δ1-82 (6.4 mg/ml) wild type in 60% monoolein cubic phases. Crystals were grown in buffer solutions containing screens listed in Table 15.

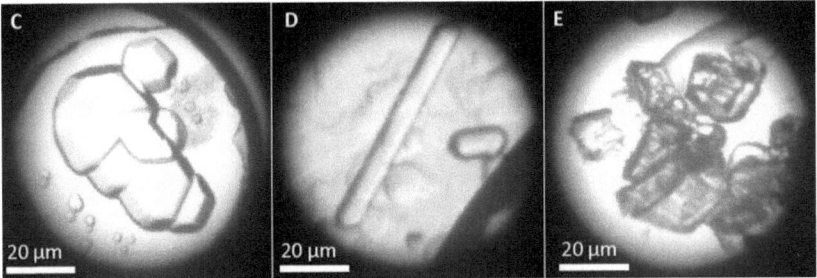

Figure 3.31: Crystals of hTom40AΔ1-82 K107L/H117L/H220L (4 mg/ml) in 60% monoolein cubic phases. Crystals were grown in buffer solutions containing screens listed in Table 15.

# 4 Discussion

The process of protein translocation through the TOM complex has been studied for decades, and though intensive insights have been gained on the fields of structure and the process of translocation, some mysteries are still waiting to be unraveled (Mokranjac and Neupert 2009). Of special interest are the mass and the stoichiometry of the complex, as well as the structure of the main component Tom40. In this work I have addressed these issues from different perspectives.

## 4.1 Mass of the TOM core complex from *N. crassa*

The mass of the TOM complex is a matter discussed extensively for a long time as several analyses methods revealed different results (Table 16). Size exclusion chromatography (SEC) and blue native gel electrophoresis (BN-PAGE) indicated a total molecular mass of TOM core and holo complex in detergent solution between 230 and 500 kDa (Künkele et al. 1998; Rapaport et al. 1998; Meisinger et al. 2001; Werhahn et al. 2003). The predicted masses were determined with purified TOM complex and always included the mass of the surrounding detergents. The mass of detergent micelles can be subtracted but it is not clear whether the micelle mass corresponds to the amount of detergent molecules surrounding the protein. Detergents can make up to 90 kDa for DDM micelles and it is still not clear if this mass could be even higher when bound to a high molecular mass complex like the TOM core complex. Another limitation is the calibration for methods like SEC or BN-PAGE, as calibration proteins, used with these methods, are usually soluble and not in detergent micelles. In addition, BN-PAGE allows only a rough size determination as the resolution in polyacrylamide-gels is not very high (Hunte 2003).

In this work an innovative method has been applied to determine the mass of the TOM core complex. Laser induced liquid bead ion desorption (LILBID) coupled with mass spectrometry (Morgner et al. 2008; Hoffmann et al. 2010; Sokolova et al. 2010) turned out to be a powerful tool to analyze the TOM complex. To this date, this technique has been successfully applied to several membrane protein complexes such as cytochrome bc1 complex and cytochrome C oxidase of the soil bacterium *Paracoccus denitrificans* (Morgner et al. 2007), the c-ring of the $F_0/F_1$-ATP synthase of various alkaliphilic bacteria (Meier et al. 2007), and the mitochondrial respiratory chain complex I of the yeast *Yarrowia lipolytica* (Morgner et al. 2008). These high-molecular mass complexes, with masses in the range of the TOM complex, have been successfully studied with LILBID without detergents attached. LILBID is not an in-vivo method, but its advantages in precision regarding subunit determination made it an ideal method to analyze the subunit composition and stoichiometry of the TOM complex.

The detergent DDM was already used for several stoichiometry analyses of membrane proteins (Morgner et al. 2007). As the TOM complex is stable in DDM (a non-ionic detergent which is needed for LILBID measurements) it sounded promising to test this method on the TOM complex. The main advantage of the exact mass determination with LILBID is that the detergent is completely removed during the laser excitation.

Interestingly, the resulting molecular mass of 170 kDa for the TOM core complex determined by LILBID-MS (Figure 3.4 B, C) was about 2.5-fold smaller than that of core complex deduced by non-denaturing gel electrophoresis and SEC (Ahting et al. 1999). However, the limitations of these methods have been mentioned above. The primary reason for this discrepancy could be the mass variability of the complex regarding the small subunits as indicated in Figure 3.4.

Another promising method to determine the exact mass of proteins is analytical ultracentrifugation. Even for membrane proteins this method is applicable as long as a suitable detergent is used to keep the protein in solution. In this case, the detergent polyoxyethylene could be suitable as it has the same density as $H_2O$ and can be subtracted easily from the background intensity. However, TOM core complex is not stable in oPOE, so this approach has been addressed for TOM core complex in DDM and revealed a total mass of 370 kDa (Nußberger, unpublished data).

Other attempts to determine the mass of the TOM complex have been done with scanning transmission electron microscopy (STEM) and revealed a mass for TOM core complex of ~170 kDa (Nußberger, unpublished data). These findings are in good accordance with the data presented here. Even in accordance with findings from STEM measurements, the results of the TOM complex mass determination by LILBID are lower than the corresponding results determined with classical methods such as SEC or BN-PAGE in previous studies. Even the low laser intensities used in LILBID, might destroy protein-protein interaction. It is further possible that the mass of 170 kDa only represents a small stable part of the native TOM complex. It needs to be shown whether two or more 170 kDa complexes form a super-complex. The mass of this super-complex could then correspond to findings from previous mass determination approaches. Nevertheless, the complex as described in this thesis with a size of 170 kDa represents a stable conformation and indicates a plausible structural composition of subunits.

The masses determined for TOM holo and core complex using classical methods like SEC or BN-PAGE never gave plausible answers to the subunit composition (Dekker et al. 1998; Ahting et al. 1999; Model et al. 2008). With masses of 300 kDa or more, it was almost impossible to calculate the stoichiometry of the other subunits present in the complex with single subunit masses ranging from 5 to 20 kDa. This problem has been solved with LILBID and will be discussed in the following.

## 4.2 Stoichiometry of the TOM complex

Although the subunit compositions of TOM complexes from fungi, mammals and plants are remarkably similar, their individual subunit stoichiometries are still a matter of controversy (Ahting et al. 1999; Schmitt et al. 2005). This issue has been addressed in previous studies using BN-PAGE, chemical cross-linking of individual subunits, or cryo-electron microscopy (Table 16). However, the total number of molecules present in the TOM core complex remained unknown.

Several methods to evaluate the stoichiometry of a large multiprotein complex are known. Some address the relative amounts of subunits while others can reveal the absolute amount in a complex. The relative amount of subunits to each other can be analyzed with radioactive labeling, which e.g. has determined the stoichiometry of the bacterial ribosome (Tal et al. 1990). Other attempts to determine the stoichiometry of large protein complexes could be done with concatamers of tryptic peptides of the subunits proteins, which, quantified with mass spectrometry, reveal the amounts of subunits in a complex (Pratt et al. 2006). With Fluorescence labeling the absolute protein stoichiometry of MotB, a protein complex with in the flagellar motor of *E. coli*, was determined by labeling MotB molecules with green fluorescent protein and stepwise photobleaching of single GFP molecules (Leake et al. 2006)

Here, the application of LILBID-MS to analyze the stoichiometry of purified TOM core complex unraveled the subunit composition in the complex. As mentioned above, the highest mass for the TOM complex was found to be ~170 kDa, determined by applying low laser intensities for the ion desorption. The peaks in the mass spectrometry did not show a single peak at a mass of 170 kDa, but ranged around this mass with peak differences of ~6 kDa. This indicates that the complex contains a stable number of its main component Tom40 and the receptor Tom22. In addition, these two proteins are flanked by a variable amount of small TOM-proteins, which have an average mass of 6 kDa each. Mass spectra obtained using higher laser intensity show that subcomplexes arose from the complex which contained 2 x Tom40 proteins and 2 x Tom22 as well as a variable amount of small Tom proteins. The rather fluctuating amount of small Tom proteins indicates that these are not very tightly attached to the rest of the complex. This shows the dynamic of the TOM machinery where some subunits might be associated only transiently.

Table 16: Publications about masses and stoichiometry of the TOM complex

| Organism | Method | Complex | Mass [kDa] | Stoichiometry | Author |
|---|---|---|---|---|---|
| S. cerevisiae | Radiolabelling, immuno-precipitation | CC | - | Tom40:Tim23:Tim22 (5:1:0.22) | (Sirrenberg et al. 1997) |
| A. thaliana | BN-PAGE | CC | 230 | - | (Jänsch et al. 1998) |
| N. crassa | Phospho imaging | HC | 504 | Tom70:Tom40:Tom22:Tom20 (1.5:8:3.1:2) | (Künkele et al. 1998) |
| N. crassa | SEC | HC | 450 | | (Rapaport et al. 1998) |
| S. cerevisiae | BN-PAGE, immunoblotting | HC | 400 | 4-6xTom40, 3-6xTom22, 6-12xsmToms | (Dekker et al. 1998) |
| N. crassa | EM, SEC, BN-PAGE | CC | 410 | 6-8xTom40 | (Ahting et al. 1999) |
| N. crassa | BN-PAGE | HC | 440 | | (Rapaport et al. 2001) |
| S. cerevisiae | BN-PAGE | CC (+Tom20) | 400 (500-600) | 2-3xTom40 | (Meisinger et al. 2001) |
| A. thaliana | BN-PAGE, Immunoblotting | CC | 230 | - | (Werhahn et al. 2001) |
| A. castellanii | BN-PAGE, ESI-MS, MALDI-MS | HC | ~500 | - | (Wojtkowska et al. 2005) |
| S. cerevisiae | Cryo-EM | CC + Tom20 | 550 | 3xTom40, 3xTom22, 1-4xTom20 | (Model et al. 2008) |
| S. cerevisiae | BN-PAGE, autoradiography | HC | 440 | - | (Becker et al. 2008) |
| N. crassa | LILBID | CC | 170 | 2xTom40, 2xTom22, 6-10 smToms | (Mager et al. 2010) |

CC: TOM core complex, HC: TOM holo complex

The oligomeric organization of Tom40 in the TOM complex was analyzed previously with chemical crosslinking, SDS-PAGE and immunostaining with antibodies against Tom40 (Rapaport et al. 1998). It revealed that Tom40 forms a homo-oligomeric assembly that persists under conditions that lead to the dissociation of the receptor components from Tom40. It further indicated that Tom40 is organized as a dimer which then forms a larger structural assembly together with the other subunits of the TOM machinery. The dimerization of Tom40 stands in line with findings presented in here about the oligomerization state of Tom40. Additionally, the dimerization of another β-barrel has been reported for VDAC, an ancestrally related protein to Tom40 (Szabo and Zoratti 1993; Keinan et al. 2010).

Given that a single Tom40 protein forms the pore through which mitochondrial preproteins cross outer mitochondrial membranes (Zeth 2010), our data suggest that the two pore structure of the TOM core complex, as previously determined by electron microscope tomography (Ahting et al. 1999), is represented by a dimer of Tom40. Two Tom22 receptor proteins and up to ten small Tom proteins may associate with this dimer to form the 170 kDa complex.

### 4.3 Structure of TOM core complex

A first structural view of the multi-subunit core complex of *N. crassa* and *S. cerevisiae* was gained from electron microscopy and single particle image analysis (Ahting et al. 1999; Ahting et al. 2001; Model et al. 2002; Model et al. 2002; Model et al. 2008). Electron microscopy studies on the TOM core complex have revealed a twin-pore structure with a pore diameter of 20 Å (Ahting et al. 1999). TOM complex has been crystallized previously, but due to its subunit heterogeneity resulted only in irregular ordered crystals and structure determination at high resolution had not been possible yet (personal communication, Nußberger). The inhomogeneity of subunits might be the reason for the difficulties during crystallization.

Besides the stoichiometry of the TOM complex, LILBID measurements also gave insights into its structural configuration. With medium laser intensities stable subcomplexes have been extracted from the complex (see Figure 3.3), and gave hints about the interaction between subunits according to the presence of certain protein arrangements.

It is remarkable that a subcomplex of 2 x Tom40 has been identified while a subcomplex of 2 x Tom22 was not detectable. This implies that two Tom40 molecules are likely be attached to each other, which is not the case for two molecules of the receptor Tom22. A subcomplex of 1 x Tom40 and 1 x Tom22 was detected in spectra recorded at medium laser intensities (Figure 3.3) claiming a stable binding between Tom40 and Tom22. This stands in line with findings in earlier publications showing a highly stable Tom40-Tom22 core structure under alkaline treatment (Meisinger et al. 2001). In addition, peaks corresponding to combinations of this Tom40-Tom22 binding, namely 2 x Tom40 and 1 x Tom22, 1 x Tom40 and 2 x Tom22 or 2 x Tom40 and 2 x Tom22, are present in the spectrum. With this information about the possible combinations in the core complex I propose a structural model stating a double pore of two Tom40 molecules with two Tom22 proteins attached, which are not located next to each other (Figure 4.1).

In the mitochondrial outer membrane the small Tom proteins Tom5, Tom6 and Tom7 may fill the spaces generated by the association of Tom40 and Tom22. Subcomplexes containing small Tom proteins can consist of 1 x Tom40 and 1-2 x smTom proteins. Even though the resolution of the LILBID mass spectrum under is ~ 1 kDa, it does not allow to differentiate between 3 x smTom proteins (~ 18 kDa) or 1 x Tom22 (17.8 kDa). A composition of 1 x Tom40 and 3 x smToms might be possible but the sharpness of the peak at 55.7 m z$^{-1}$ refers to a subcomplex of 1 x Tom40 and 1 x Tom22 as indicated in Figure 3.4. It remains to mention that a subcomplex containing only two or more small Tom proteins, lacking Tom22 or Tom40, has not been detected, indicating that the small Tom proteins need these larger Tom proteins for assembly or that the interactions between two small Tom proteins alone are too weak to be detected with LILBID.

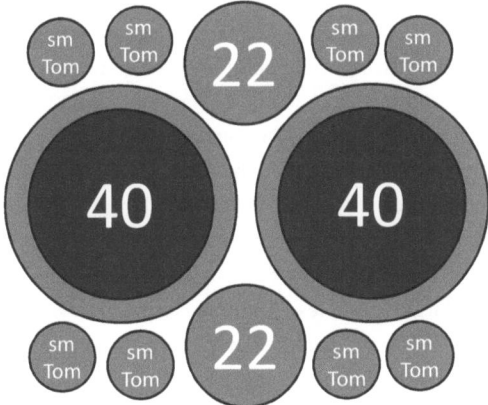

Figure 4.1: Structural arrangement of subunits proposed for TOM core complex based on data from LILBID mass spectrometry. A complex of 170 kDa has been identified consisting of 2 x Tom40, 2 x Tom22 and a variable amount of small Tom proteins (smTom).

As an interaction between two Tom22 proteins has not been detected, a subcomplex containing 1 x Tom22 and 3 x smTom can also be excluded. This may indicate that Tom22 has two binding sites for small Tom proteins. In spectra recorded under high laser intensities, when most of the complex disintegrates into its subunits, a subcomplex containing Tom22 and Tom6 was still detectable indicating a tight interaction between the two proteins. Since the resolution at medium laser intensities does not allow an exact distinction of the small Tom proteins, it is possible that the small Tom proteins attached to Tom22 are also Tom5 or Tom7. Previous crosslinking studies indicated the attachment of the small Tom proteins Tom6 and Tom7 to Tom40, and a tight interaction between Tom22 and Tom6 (Dembowski et al. 2001).

Binding-site residues in proteins are known to be more conserved than the rest of the surface (Konc and Janezic 2007), and finding local surface similarities by comparing a protein to its interaction partners could reveal the location of binding sites on this protein. To identify possible binding sites between Tom22 and Tom6, the protein sequence was analyzed by determining the conservation of residues (Ashkenazy et al. 2010). This analysis revealed conserved amino acids in the residues 31-52 in the C-terminal part of Tom6 and in residues 80-120 in Tom22. Mutations in theses residues of Tom6 and Tom22 and subsequent LILBID analysis may provide direct evidence that these amino acid residues are indeed involved in the interaction of Tom22 and Tom6.

Single-particle electron microscopy analysis of *N. crassa* and yeast TOM holo complex led to different views with two or three stain-filled centers resembling channels (Model et al. 2002). The TOM complex from mutant yeast, selectively lacking Tom20, showed only particles with two pore structure. These findings are in good accordance with our LILBID data indicating a complex lacking Tom20. From mutant yeast mitochondria lacking Tom22, BN-PAGE reveals a ~80 kDa subcomplex that consists only of Tom40. It has been stated that the two receptors Tom22 and Tom20 are required for the organization of Tom40 dimers into larger TOM structures (Model et al. 2002). Even though the structure of the TOM complex that we propose based on the LILBID data comprises only two pores, it remains possible that the binding of Tom20 results in a larger complex. The measurements with LILBID have been performed with the TOM core complex while analysis of the holo complex was postponed due to lack of the receptors Tom20 and Tom70 in a suitable detergent for LILBID measurements. It is possible that a three-pore complex is only stable when Tom20 is still present and that the complex disintegrates in to smaller subcomplexes upon removal of Tom20 (Model et al. 2008). A structural model for TOM core complex taking the findings from LILBID measurements and previous data into account is shown in Figure 4.1.

An earlier study about the pore properties of Tom40 and its role in the complex claimed that a single Tom40 protein would not be sufficient to generate a pore with a diameter of 20 Å (Ahting et al. 2001). However, these findings were based on the assumption that Tom40 contains only 8-10 β-strands which would not be sufficient to build a pore with the defined diameter. Recent analysis of the secondary structure of Tom40 indicates a pore composed of 19 β-strands (Figure 1.3). This is not surprising, since the algorithms, on which secondary structure predictions are based, are updated frequently including the latest scientific findings (Cole et al. 2008). Considering these theoretical data and the structural data obtained in this work I propose that each pore in the TOM complex is formed by a single Tom40 protein.

## 4.4 Recombinant expression of different Tom40 isoforms

The structure of the TOM core complex of *N. crassa* had successfully been studied with LILBID. However, crystallization attempts with the whole complex did not yield further results on the structure. Therefore, the focus was shifted towards the main protein in the complex, Tom40. The approach to purify Tom40 from *N. crassa* gave pure protein, but the yield was not sufficient for crystallization attempts. We decided to express Tom40 recombinantly to obtain higher amounts of pure protein than it has been gained with native purification from mitochondria.

When studying mammalian Tom40s it is remarkable that, unlike in yeast and fungus, there are several isoforms. This is not only the case in mammals but also occurs in *A. thaliana*, which comprises two isoforms as well (Macasev et al. 2000; Humphries et al. 2005). The second mammalian Tom40 isoform, termed Tom40B, has first been described in rat (Kinoshita et al. 2007). In this study I structurally characterized the two human isoforms of Tom40 for the first time. They share the common structural feature of all known Tom40s which is a β-barrel. Both human isoforms share a sequence similarity of 65 % in this part. The main difference between both isoforms lies in the N-terminal elongation where the amino acid sequence has undergone several mutations, since both isoforms descended from a common ancestor protein.

Regardless of the N-terminus, both isoforms are functional components of the TOM complex. They are both capable of binding preproteins and translocating them into the IMS. The appearance of the two isoforms seems to be ubiquitous in all tissues. Therefore it is obvious that both Tom40 isoforms are present in the same cells and maybe even side by side in the same complex. However, in testis a lower concentration of Tom40B has been assessed. Moreover, some Tom20 isoforms in *D. melanogaster* showed selective appearance in testis (Hwa et al. 2004). It is suggested that certain isoforms of the TOM complex might play a role in spermatogenesis. A second Tom40 isoform might give way to an alternate targeting or translocation pathway in certain tissues to increase the capacity of mitochondrial import.

The N-terminal elongation in human Tom40 isoform A is enriched in prolines. Prolines have the tendency to form unstructured amino acid chains but proline-rich regions (PRRs) can also form polyproline-helices as present in collagen or rarely in globular proteins (Gratzer et al. 1963; Adzhubei and Sternberg 1993).

The PRR in the N-terminus is unique for mammalian Tom40 and exclusively found in isoform A. In other Tom40s from fungus, yeast or other mammalian isoforms no accumulation of prolines is found. It is still not clear why mammalian and plant TOM complexes comprise several isoforms of their channel protein. Their function as protein transporter must be similar as they share most sequence similarity in the β-barrel part. The N-terminal elongation exhibits more variability and is also different in length between isoforms and among species. The specific function of the N-terminus still remains unclear. It is assumed that is assists in protein translocation or the binding of preproteins but it is not necessarily required as shown in several studies where a truncated Tom40 protein is still capable to transport proteins. Even the folding of protein cannot be fully dependent on the N-terminus as truncated Tom40 mutants show proper folding and channel activity (Suzuki et al. 2004).

Proteins with proline rich regions are found in both prokaryotes and eukaryotes (Williamson 1994). Some outer membrane proteins of cyanobacteria like Omp85 have a proline-rich region of variable length in their N-terminus (Arnold et al. 2010). Prolines can be involved in protein binding as reported for the epithelial sodium channel ENaC (Kanelis et al. 2000). The mechanism of protein-protein binding cannot be exclusively carried out by the proline-rich N-terminus in hTom40A as protein translocation can be mediated by other isoforms as well. Also other parts of Tom40 are responsible for protein binding partly with the assistance of other Tom proteins.

The truncation of the N-terminus done in this work was mainly based on the fact that proline rich regions are unfavorable for crystallization. Long regions with proline repeats might hinder a specific folding and therefore might inhibit the formation of a unit cell (Levitt 1981). As the function of the N-terminus in mammalian Tom40 is not clear, a truncation seemed opportune to focus on the structure and function of the β-barrel.

## 4.5 Secondary structure determination: applications and limitations

After hTom40A and B were expressed recombinantly in *E. coli* and refolded in detergent, their folding state was initially assessed by CD spectroscopy. The data revealed that the protein was properly folded and indicated a high amount of β-sheets.

Exact prediction of secondary structure content with Software as CDpro is limited for globular proteins, especially membrane proteins with a high β-sheet content. Calculations done with CDpro for spectra of Tom40 always revealed much lower β-sheet content than from calculations of FTIR data or secondary structure predictions. This discrepancy between secondary structure calculations from CD and predictions made with Psipred had previously been shown with VDAC when the structure was solved (see Table 11). However, CD spectroscopy measurements done in this work were only used to assess the quality of the protein folding. Requirements for an accurate secondary structure calculation of CD data were not needed. The method is, however, very reliable for monitoring changes in the conformation of proteins under different conditions like denaturation studies or unfolding experiments. In this work it has been shown that Tom40 unfolds in a two-step process, with the α-helix denaturating prior to the β-barrel. This was indicated by changes in secondary structure content of CD spectra recorded during thermal unfolding.

Secondary structure of β-barrels can be measured more precisely with FTIR. However, the algorithm for the secondary structure prediction in this work was not optimized for β-barrel proteins as it is based on FTIR spectra of proteins with known structure (Bruker Protein Spectra Library). This library does not include typical β-barrel proteins so improvement on the structure prediction could be made by including secondary structure contents of proteins solved in the past years like VDAC1 or bacterial porins (Pauptit et al. 1991; Cowan et al. 1992; Bayrhuber et al. 2008; Ujwal et al. 2008).

Comparison with previous FTIR spectra recorded for Tom40 and VDAC from *N. crassa* revealed remarkable issues (Ahting et al. 2001). The spectrum for NcTom40 shows a maximum at 1650 cm-1 which indicates a high α-helical content. This stands in contrast to spectra recorded for hTom40 shown in this work (Figure 3.13, Figure 3.22). More similarity can be found in spectrum recorded for VDAC from *N. crassa* in Ahting et al. when compared with spectra of hTom40 in this work. Both spectra show a comparable high β-sheet content. However, it has to be mentioned that the measurements from Ahting et al. were performed with dried protein samples on a Germanium crystal which might influence the folding state of the protein. In this work I recorded the FTIR spectra with proteins in solution which might represent a more native state.

## 4.6 Physiological properties of Tom40

To examine the channel properties of recombinant Tom40, both isoforms were reconstituted in black lipid membranes and their ability to conduct an ion flow was assessed. Even though the insertion events of a single channel in the membrane were rare, the channels' ability to conduct an ion flow has been proven, indicating the successful refolding to a functional channel protein. A single channel conductance was hard to determine due to the low frequency of channel insertions. In the rare cases when channels inserted in the membrane the voltage dependence was assessed. It was remarkable that even at high voltages up to ± 160 mV the channel did not show any or just remote closure. In contrast, other Tom40 proteins showed a two-state channel conductance, which occurred at voltages higher than 100 mV (Becker et al. 2005). It is possible that this gating process in other Tom40s is dependent on the N-terminal helix which might be facing the lumen of the channel and influence the conductance state. The results here also stand in contrast to channel characteristics of the structurally related channel VDAC which closes down to 50% conductance under voltage applications of 50 mV or more (Engelhardt et al. 2007). Even though the helix in the truncated versions of human Tom40A and B used in this work is still present, the truncations of the first 82 and 29 amino residues in hTom40A and B, respectively, can be responsible for the loss of gating ability.

## 4.7 Protein aggregation: True or not?

Recombinant expression and refolding of membrane proteins always bear the risk of oligomerized proteins or aggregation products. With size exclusion chromatography (SEC) the oligomerization state of the protein can be determined by its particle size. When using SEC in this work the protein was always eluted in the void volume of any column used which would imply that the protein formed high oligomers. This has been noticed using columns with a separation range of 1-100 kDa like Superose12 or Superdex75 but also using columns with a larger separation range from 5-5000 kDa like Superose6. The question arose whether this can be true for a protein with a predicted mass of 32 kDa to form oligomers exceeding 5000 kDa. To evaluate this matter I performed SEC with hTom40 in various detergents resulting every time in the same elution profile with the protein eluting in the void volume. Some chromatograms showed a second elution peak corresponding to a protein size approaching the actual mass of Tom40 with ~40 kDa. This appearance varied according to the concentration assuming that a lower protein concentration might promote the formation of protein monomers.

However, there is quite some evidence that SEC is not the method of choice for size determination of membrane proteins in detergent solution (Ma and Xia 2008). The detergents monomers and micelles can interfere with the gel matrix, blocking the pores or changing the matrix environment. It is ascertained that the protein radius and mass is massively falsified in detergent solution. Some detergents like DDM have a fixed micelle size in solution which might still vary in when surrounding a protein. For other detergents like LDAO no definite micelle size can be assessed (Arnold and Linke 2008). Also the operation of SEC under high pressure, which can be up to 30 bar for some analytical SEC columns, might influence the aggregation state of the protein.

To exclude any influence of detergent on the oligomerization state, recombinant Tom40 in 6 M guanidine-hydrochloride was applied to SEC as well and surprisingly showed no difference in the elution profile. This indicates that even in high chaotropic solutions as 6 M guanidine-hydrochloride, which was proven to unfold the proteins efficiently, no monomeric Tom40 was detectable with SEC.

A mass determination by SEC is usually accompanied by an elution profile of reference proteins with known mass and their elution peak is compared to the analyzed protein sample. Comparing proteins with greatly different hydrophobicity characteristics with a method like SEC is prone to error and therefore not suitable for an exact size determination (Welling-Wester et al. 1988; Grisshammer and Buchanan 2006).

A promising attempt to examine the oligomerization state of Tom40 in this work was made with dynamic light scattering (DLS). This method is applicable for soluble proteins as well as for proteins in detergent solution. DLS is an established technique in the field of protein crystallography as it measures the hydrodynamic sizes, polydispersities and aggregation effects of protein samples (Miyatake et al. 1999; Proteau et al. 2010). Its accuracy in determining the mass of the protein analyte is not very high, but by calculating the hydrodynamic radii of particles in solution the aggregation state of proteins can be determined precisely. Therefore, I applied DLS to refolded protein samples after SEC chromatography revealed puzzling results. DLS indicated monodisperse particles in the protein samples.

## 4.8 Stabilization of the β-barrel

There is evidence that in the TOM-complex Tom40 is present as a dimer. Tom40 is embedded in the complex and surrounded by other components of the TOM complex which may shield the β-barrel from the membrane environment.

Several other β-barrel membrane proteins, like NalP or ScrY, which are not part of a complex, comprise an additional stabilizing element. This element may compensate for an otherwise unfavorable energetic composition. The stabilization can either be achieved by a structural element like a α-helix, which can stabilize the barrel from the outside as out-clamp, or from the inside as in-plug. Another possibility to stabilize barrels is the formation of oligomers, which shields unstable regions from the environment (Naveed et al. 2009). Tom40 may comprise two stabilizing elements. The N-terminus of Tom40, forming partly a α-helix, may stabilize the protein as out-clamp or as in-plug. In addition, Tom40 might form oligomers to stabilize itself.

Calculating the energy levels of the 19 β-strands revealed 3 strands to be less stable than the others. A closer look in these strands revealed one amino acid in each strand to be responsible for this instability. After these amino acids had been identified, I replaced them by hydrophobic amino acids and therefore adapted the energy level to the surrounding amino acids. The theoretical calculation and the subsequent mutagenesis resulted in a remarkably more stable mutant Tom40. In contrast to the wild type, the mutant protein showed significant advantages concerning thermal and chemical stability.

By incubation with the chemical crosslinker glutaraldehyde the mutant protein showed reduced oligomerization in comparison with the wild type protein. Conclusions from crosslinking experiments could be made, as the crosslinker only connects proteins which are located close to each other in a complex or in a protein solution. With the result showing a reduced tendency of the mutated Tom40 to form oligomers, it can be stated that the wild type protein predominantly forms oligomers in solution to shield its weakly stable β-strands by forming dimers or trimers. The mutation of the destabilizing amino acids successfully showed the improvement of the oligomerization state towards the monomer. The increased stability provides a promising base for further structural investigations on Tom40.

## 4.9 Crystallization

Since it is impossible to fully understand protein functionality without knowing its exact molecular nature, it is essential to get structural information about the protein first. Structural insights into protein architecture can be achieved by nuclear magnetic resonance (NMR) or x-ray crystallography. NMR is a convenient method but is yet only applicable to proteins smaller than ~ 30 kDa, also it is more complex to apply to membrane proteins (Sheehan 2009). As Tom40 is expected to be structurally related to VDAC, whose crystal structure was solved, this method seemed appropriate for Tom40 as well (Bayrhuber et al. 2008). First attempts to crystallize Tom40 were successful but the obtained crystals from protein in detergent solution did not show an adequate resolution to determine the structure. In the crystallization trials LDAO was used to solubilize Tom40 as several crystal structures have been solved using this detergent (Hiller et al. 2008). LDAO is, along with other detergents like DDM, β-OG, and oPOE, one of the most successful detergents for membrane protein crystallization (Aartsma and Matysik 2008). However, it may be possible that high LDAO concentrations so far prevented the formation of a densely packed protein crystal in the crystallization trials carried out in this study.

One of the frequent challenges in crystallization is the solubility of the protein at high protein concentrations. Membrane proteins are most stable in their native bilayer environment but mimicking such environments *in-vitro* is a tough task. This problem can be solved doing the crystallization in bicelles (Faham and Bowie 2002). By mixing of protein and bicelles, the protein will insert into the lamellar bicelle discs, in which it is expected to be more stable and therefore more suitable for crystallization (Faham et al. 2005). From a practical point of view, crystallization in bicelles is similar to that in detergent micelles. Both methods can be set up by robot-assisted pipetting and therefore enables a high-throughput screening with different additive solutions. This crystallization method was successfully applied to solve the structure of mouse VDAC1 (Ujwal et al. 2008). It is intuitive that the organization of the protein in a lipid environment is closest to its native state in the membrane. The crystallization of Tom40 on the basis of the findings in this work is a promising and still on-going project.

## 4.10 Outlook

The structural composition of *N. crassa* TOM complex has been approached previously with x-ray crystallography, but did not result in high resolution crystal formation so far. As shown in this work, the inhomogeneity regarding the number of the small Tom proteins per complex might be a reason for this failure. So it seemed more promising to approach the overall structure of the complex by starting with the single subunits. In this work significant improvement has been made on the way to crystallize the main component of the TOM complex, Tom40. The protein has been stabilized and therefore provides a promising base to obtain better ordered crystals compared to wild type Tom40. This matter will be addressed in the near future and already revealed initial success.

Another challenging task could be the behavior of human Tom40 K107L/H117L/H220L in the TOM complex. It might be interesting to determine how the protein integrates in the complex and how the mutations influence the complex composition *in vivo* in mitochondrial membranes.

In this work the *in vitro* stoichiometry of the TOM complex was successfully analyzed with LILBID. The complex had to be purified in detergent solution and might have a different composition *in vivo*. To investigate the stoichiometry in natural lipid environment, the complex could be isolated in its membrane after single subunits were fused to fluorescent protein labels or organic dyes. With a sufficient low density of the complex in the membrane, it would be possible to count individual molecules by single-molecule fluorescence bleaching, as it has been shown previously for the voltage-gated potassium channel (Nakajo et al. 2010). With this approach the purification with detergents would become redundant and the amount of subunits in the TOM complex could be assessed closer to their natural state.

# 5 Bibliography

Aartsma, T. J. and J. Matysik (2008). "Biophysical techniques in photosynthesis." **Vol. 2**: 501.
Abe, Y., et al. (2000). "Structural basis of presequence recognition by the mitochondrial protein import receptor Tom20." Cell **100**: 551-560.
Adzhubei, A. A. and M. J. Sternberg (1993). "Left-handed polyproline II helices commonly occur in globular proteins." J Mol Biol **229**(2): 472-493.
Ahting, U., et al. (2001). "Tom40, the pore-forming component of the protein-conducting TOM channel in the outer membrane of mitochondria." J Cell Biol **153**(6): 1151-1160.
Ahting, U., et al. (1999). "The TOM core complex: the general protein import pore of the outer membrane of mitochondria." J Cell Biol **147**(5): 959-968.
Arnold, T. and D. Linke (2008). "The Use of Detergents to Purify Membrane Proteins." Current Protocols in Protein Science UNIT 4.8.
Arnold, T., et al. (2007). "Gene duplication of the eight-stranded beta-barrel OmpX produces a functional pore: a scenario for the evolution of transmembrane beta-barrels." J Mol Biol **366**(4): 1174-1184.
Arnold, T., et al. (2010). "Omp85 from the thermophilic cyanobacterium Thermosynechococcus elongatus differs from proteobacterial Omp85 in structure and domain composition." J Biol Chem **285**(23): 18003-18015.
Ashkenazy, H., et al. (2010). "ConSurf 2010: calculating evolutionary conservation in sequence and structure of proteins and nucleic acids." Nucleic Acids Res **38**(Web Server issue): W529-533.
Baker, K. P., et al. (1990). "A yeast mitochondrial outer membrane protein essential for protein import and cell viability." Nature **348**(6302): 605-609.
Bauer, M. F., et al. (1996). "Role of Tim23 as voltage sensor and presequence receptor in protein import into mitochondria." Cell **87**(1): 33-41.
Bauerschmitt, H., et al. (2010). "Ribosome-binding proteins Mdm38 and Mba1 display overlapping functions for regulation of mitochondrial translation." Mol Biol Cell **21**(12): 1937-1944.
Bayrhuber, M., et al. (2008). "Structure of the human voltage-dependent anion channel." Proc Natl Acad Sci U S A **105**(40): 15370-15375.
Becker, L., et al. (2005). "Preprotein translocase of the outer mitochondrial membrane: reconstituted Tom40 forms a characteristic TOM pore." J Mol Biol **353**(5): 1011-1020.
Becker, T., et al. (2008). "Biogenesis of the mitochondrial TOM complex: Mim1 promotes insertion and assembly of signal-anchored receptors." J Biol Chem **283**(1): 120-127.
Becker, T., et al. (2010). "Biogenesis of Mitochondria: Dual Role of Tom7 in Modulating Assembly of the Preprotein Translocase of the Outer Membrane." J Mol Biol.
Bolender, N., et al. (2008). "Multiple pathways for sorting mitochondrial precursor proteins." EMBO Rep **9**(1): 42-49.
Bolliger, L., et al. (1995). "Acidic receptor domains on both sides of the outer membrane mediate translocation of precursor proteins into yeast mitochondria." Embo J **14**(24): 6318-6326.
Bradford, M. M. (1976). "A rapid and sensitive method for the quantitation of microgram quantities of protein utilizing the principle of protein-dye binding." Anal Biochem **72**: 248-254.
Brix, J., et al. (1997). "Differential recognition of preproteins by the purified cytosolic domains of the mitochondrial import receptors Tom20, Tom22, and Tom70." J Biol Chem **272**(33): 20730-20735.
Caffrey, M. (2003). "Membrane protein crystallization." J Struct Biol **142**(1): 108-132.
Caffrey, M. and V. Cherezov (2009). "Crystallizing membrane proteins using lipidic mesophases." Nat Protoc **4**(5): 706-731.
Chacinska, A., et al. (2009). "Importing Mitochondrial Proteins: Machineries and Mechanisms." Cell **138**(4): 628-644.
Chacinska, A., et al. (2005). "Mitochondrial presequence translocase: switching between TOM tethering and motor recruitment involves Tim21 and Tim17." Cell **120**(6): 817-829.

Chan, N. C., et al. (2006). "The C-terminal TPR domain of Tom70 defines a family of mitochondrial protein import receptors found only in animals and fungi." J Mol Biol **358**(4): 1010-1022.
Chan, N. C. and T. Lithgow (2008). "The peripheral membrane subunits of the SAM complex function codependently in mitochondrial outer membrane biogenesis." Mol Biol Cell **19**(1): 126-136.
Chew, O., et al. (2004). "A plant outer mitochondrial membrane protein with high amino acid sequence identity to a chloroplast protein import receptor." FEBS Lett **557**(1-3): 109-114.
Cole, C., et al. (2008). "The Jpred 3 secondary structure prediction server." Nucleic Acids Res **36**(Web Server issue): W197-201.
Colin, J., et al. (2009). "The mitochondrial TOM complex modulates bax-induced apoptosis in Drosophila." Biochem Biophys Res Commun **379**(4): 939-943.
Colombini, M. (1979). "A candidate for the permeability pathway of the outer mitochondrial membrane." Nature **279**(5714): 643-645.
Cowan, S. W., et al. (1992). "Crystal structures explain functional properties of two E. coli porins." Nature **358**(6389): 727-733.
Dagley, M. J., et al. (2009). "The protein import channel in the outer mitosomal membrane of Giardia intestinalis." Mol Biol Evol **26**(9): 1941-1947.
Das, G. and S. Matile (2001). "Topological diversity of artificial beta-barrels in water." Chirality **13**(3): 170-176.
Dekker, P. J., et al. (1998). "Preprotein translocase of the outer mitochondrial membrane: molecular dissection and assembly of the general import pore complex." Mol Cell Biol **18**(11): 6515-6524.
Dembowski, M., et al. (2001). "Assembly of Tom6 and Tom7 into the TOM core complex of Neurospora crassa." J Biol Chem **276**(21): 17679-17685.
Dietmeier, K., et al. (1997). "Tom5 functionally links mitochondrial preprotein receptors to the general import pore." Nature **388**(6638): 195-200.
Dimmer, K., S. and D. Rapaport (2010). "The enigmatic role of Mim1 in mitochondrial biogenesis." Eur J Cell Biol **89**(2-3): 212-215.
Dolezal, P., et al. (2006). "Evolution of the molecular machines for protein import into mitochondria." Science **313**(5785): 314-318.
Dolezal, P., et al. (2005). "Giardia mitosomes and trichomonad hydrogenosomes share a common mode of protein targeting." Proc Natl Acad Sci U S A **102**(31): 10924-10929.
Eliyahu, E., et al. (2010). "Tom20 mediates localization of mRNAs to mitochondria in a translation-dependent manner." Mol Cell Biol **30**(1): 284-294.
Endo, T., et al. (2003). "Functional cooperation and separation of translocators in protein import into mitochondria, the double-membrane bounded organelles." J Cell Sci **116**(Pt 16): 3259-3267.
Engelhardt, H., et al. (2007). "High-level expression, refolding and probing the natural fold of the human voltage-dependent anion channel isoforms I and II." J Membr Biol **216**(2-3): 93-105.
Esaki, M., et al. (2003). "Tom40 protein import channel binds to non-native proteins and prevents their aggregation." Nat Struct Biol **10**(12): 988-994.
Esaki, M., et al. (2004). "Mitochondrial protein import. Requirement of presequence elements and tom components for precursor binding to the TOM complex." J Biol Chem **279**(44): 45701-45707.
Eskes, R., et al. (1998). "Bax-induced cytochrome C release from mitochondria is independent of the permeability transition pore but highly dependent on Mg2+ ions." J Cell Biol **143**(1): 217-224.
Fabian, H., Schultz, C.P. (2000). Encyclopedia of Analytical Chemistry. Chichester, John Wiley & Sons, Ltd.
Faham, S., et al. (2005). "Crystallization of bacteriorhodopsin from bicelle formulations at room temperature." Protein Sci **14**(3): 836-840.
Faham, S. and J. U. Bowie (2002). "Bicelle crystallization: a new method for crystallizing membrane proteins yields a monomeric bacteriorhodopsin structure." J Mol Biol **316**(1): 1-6.
Forst, D., et al. (1998). "Structure of the sucrose-specific porin ScrY from Salmonella typhimurium and its complex with sucrose." Nat Struct Biol **5**(1): 37-46.

Gasteiger, E., Hoogland C., Gattiker A., Duvaud S., Wilkins M.R., Appel R.D., Bairoch A. (2005). "Protein Identification and Analysis Tools on the ExPASy Server." The Proteomics Protocols Handbook: 571-607
Gentle, I. E., et al. (2005). "Molecular architecture and function of the Omp85 family of proteins." Mol Microbiol **58**(5): 1216-1225.
Gentle, I. E., et al. (2007). "Conserved motifs reveal details of ancestry and structure in the small TIM chaperones of the mitochondrial intermembrane space." Molecular Biology and Evolution **24**(5): 1149-1160.
Gratzer, W. B., et al. (1963). "Optical Properties of the Poly-L-proline and Collagen Helices." Biopolymers **1**(4): 319-330.
Gray, M. W. (1999). "Evolution of organellar genomes." Curr Opin Genet Dev **9**(6): 678-687.
Gray, M. W., et al. (1999). "Mitochondrial evolution." Science **283**(5407): 1476-1481.
Green, D. R. and J. C. Reed (1998). "Mitochondria and apoptosis." Science **281**(5381): 1309-1312.
Grisshammer, R. K. and S. K. Buchanan (2006). Structural biology of membrane proteins RSC Publishing.
Hannavy, K., et al. (1990). "TonB protein of Salmonella typhimurium. A model for signal transduction between membranes." J Mol Biol **216**(4): 897-910.
Herrmann, J. M. and R. Kohl (2007). "Catch me if you can! Oxidative protein trapping in the intermembrane space of mitochondria." J Cell Biol **176**(5): 559-563.
Hill, K., et al. (1998). "Tom40 forms the hydrophilic channel of the mitochondrial import pore for preproteins." Nature **395**: 516-521.
Hiller, S., et al. (2008). "Solution structure of the integral human membrane protein VDAC-1 in detergent micelles." Science **321**(5893): 1206-1210.
Hoffmann, J., et al. (2010). "Studying the stoichiometries of membrane proteins by mass spectrometry: microbial rhodopsins and a potassium ion channel." Phys Chem Chem Phys **12**(14): 3480-3485.
Hoffmann, J., et al. (2010). "ATP synthases: cellular nanomotors characterized by LILBID mass spectrometry." Phys Chem Chem Phys **12**(41): 13375-13382.
Hönlinger, A., et al. (1996). "Tom7 modulates the dynamics of the mitochondrial outer membrane translocase and plays a pathway-related role in protein import." EMBO J **15**(9): 2125-2137 issn: 0261-4189.
Huang, S., et al. (2009). "Refining the definition of plant mitochondrial presequences through analysis of sorting signals, N-terminal modifications, and cleavage motifs." Plant Physiol **150**(3): 1272-1285.
Humphries, A. D., et al. (2005). "Dissection of the mitochondrial import and assembly pathway for human Tom40." J Biol Chem **280**(12): 11535-11543.
Hunte, C. v. J., G.; Schägger, H. (2003). Membrane Protein Purification and Crystallization - A Practical Guide, Academic Press.
Huyghues-Despointes, B. M., et al. (2001). "Measuring the conformational stability of a protein by hydrogen exchange." Methods Mol Biol **168**: 69-92.
Hwa, J. J., et al. (2004). "Germ-line specific variants of components of the mitochondrial outer membrane import machinery in Drosophila." FEBS Lett **572**(1-3): 141-146.
Hwang, D. K., et al. (2007). "Tim54p connects inner membrane assembly and proteolytic pathways in the mitochondrion." J Cell Biol **178**(7): 1161-1175.
Jänsch, L., et al. (1998). "Unique composition of the preprotein translocase of the outer mitochondrial membrane from plants." J. Biol. Chem. **273**(27): 17251-17257.
Jentoft, N. (1990). "Why are proteins O-glycosylated?" Trends Biochem Sci **15**(8): 291-294.
Jia, L., et al. (2003). "Yeast Oxa1 interacts with mitochondrial ribosomes: the importance of the C-terminal region of Oxa1." Embo J **22**(24): 6438-6447.
Jones, D. T. (1999). "Protein secondary structure prediction based on position-specific scoring matrices." J Mol Biol **292**(2): 195-202.
Kanamori, T., et al. (1997). "Probing the environment along the protein import pathways in yeast mitochondria by site-specific photocrosslinking." Proc Natl Acad Sci U S A(94): 485-490.

Kanelis, V., et al. (2000). "Sequential assignment of proline-rich regions in proteins: application to modular binding domain complexes." J Biomol NMR **16**(3): 253-259.
Kassenbrock, C. K., et al. (1993). "Genetic and biochemical characterization of ISP6, a small mitochondrial outer membrane protein associated with the protein translocation complex." EMBO J **12**(8): 3023-3034 issn: 0261-4189.
Kato, H. and K. Mihara (2008). "Identification of Tom5 and Tom6 in the preprotein translocase complex of human mitochondrial outer membrane." Biochem Biophys Res Commun **369**(3): 958-963.
Keinan, N., et al. (2010). "Oligomerization of the mitochondrial protein voltage-dependent anion channel is coupled to the induction of apoptosis." Mol Cell Biol **30**(24): 5698-5709.
Khrapunov, S. (2009). "Circular dichroism spectroscopy has intrinsic limitations for protein secondary structure analysis." Anal Biochem **389**(2): 174-176.
Kiebler, M., et al. (1993). "The mitochondrial receptor complex: a central role of MOM22 in mediating preprotein transfer from receptors to the general insertion pore." Cell **74**(3): 483-492.
Kiebler, M., et al. (1990). "Identification of a mitochondrial receptor complex required for recognition and membrane insertion of precursor proteins." Nature **348**(6302): 610-616.
Kinoshita, J. Y., et al. (2007). "Identification and characterization of a new tom40 isoform, a central component of mitochondrial outer membrane translocase." J Biochem **141**(6): 897-906.
Komiya, T., et al. (1998). "Interaction of mitochondrial targeting signals with acidic receptor domains along the protein import pathway: evidence for the 'acid chain' hypothesis." EMBO-J **17**(14): 3886-3898 issn: 0261-4189.
Komiya, T., et al. (1996). "Cytoplasmic chaperones determine the targeting pathway of precursor proteins to mitochondria." Embo J **15**(2): 399-407.
Konc, J. and D. Janezic (2007). "Protein-protein binding-sites prediction by protein surface structure conservation." J Chem Inf Model **47**(3): 940-944.
Kozjak, V., et al. (2003). "An essential role of Sam50 in the protein sorting and assembly machinery of the mitochondrial outer membrane." J Biol Chem **278**(49): 48520-48523.
Künkele, K.-P., et al. (1998). "The isolated complex of the translocase of the outer membrane of mitochondria." Journal of Biological Chemistry **273**(47): 31032-31039.
Künkele, K. P., et al. (1998). "The preprotein translocation channel of the outer membrane of mitochondria." Cell **93**(6): 1009-1019 issn: 0092-8674.
Kutik, S., et al. (2008). "Dissecting membrane insertion of mitochondrial beta-barrel proteins." Cell **132**(6): 1011-1024.
Ladokhin, A. S., et al. (2000). "How to measure and analyze tryptophan fluorescence in membranes properly, and why bother?" Anal Biochem **285**(2): 235-245.
Landau, E. M. and J. P. Rosenbusch (1996). "Lipidic cubic phases: a novel concept for the crystallization of membrane proteins." Proc Natl Acad Sci U S A **93**(25): 14532-14535.
Lang, B. F., Brinkmann H., Koski L.B., Fujishima, M., Görtz, H.-D. and Burger, G. (2005). "On the origin of mitochondria and Rickettsia-related eukaryotic endosymbionts." Jpn. J. Protozool. **38**: 171-183.
Leake, M. C., et al. (2006). "Stoichiometry and turnover in single, functioning membrane protein complexes." Nature **443**(7109): 355-358.
Lepault, J., et al. (1988). "Three-dimensional reconstruction of maltoporin from electron microscopy and image processing." Embo J **7**(1): 261-268.
Levitt, M. (1981). "Effect of proline residues on protein folding." J Mol Biol **145**(1): 251-263.
Lithgow, T. (2000). "Targeting of proteins to mitochondria." FEBS Lett. **476**: 22-26.
Lueder, F. and T. Lithgow (2009). "The three domains of the mitochondrial outer membrane protein Mim1 have discrete functions in assembly of the TOM complex." FEBS Lett.
Ma, J. and D. Xia (2008). "The use of blue native PAGE in the evaluation of membrane protein aggregation states for crystallization." J Appl Crystallogr **41**(Pt 6): 1150-1160.

Macasev, D., et al. (2000). "How do plant mitochondria avoid importing chloroplast proteins? Components of the import apparatus Tom20 and Tom22 from Arabidopsis differ from their fungal counterparts." Plant Physiol. **123**: 811-816.

Macasev, D., et al. (2004). "'Tom22', an 8-kDa trans-site receptor in plants and protozoans, is a conserved feature of the TOM complex that appeared early in the evolution of eukaryotes." Mol Biol Evol **21**(8): 1557-1564.

Mager, F., et al. (2010). "LILBID-mass spectrometry of the mitochondrial preprotein translocase TOM." J Phys Condens Matter **22**(45): 454132.

Malia, T. J. and G. Wagner (2007). "NMR structural investigation of the mitochondrial outer membrane protein VDAC and its interaction with antiapoptotic Bcl-xL." Biochemistry **46**(2): 514-525.

Martinez-Caballero, S., et al. (2007). "Tim17p regulates the twin pore structure and voltage gating of the mitochondrial protein import complex TIM23." J Biol Chem **282**(6): 3584-3593.

Matouschek, A., et al. (1997). "Active unfolding of precursor proteins during mitochondrial protein import." EMBO J **16**(22): 6727-6736 issn: 0261-4189.

Mayer, A., et al. (1995). "MOM22 is a receptor for mitochondrial targeting sequences and cooperates with MOM19." EMBO J **14**(17): 4204-4211 issn: 0261-4189.

Mayer, A., et al. (1995). "Mitochondrial protein import: reversible binding of the presequence at the trans side of the outer membrane drives partial translocation and unfolding." Cell **80**(1): 127-137 issn: 0092-8674.

McGuffin, L. J., et al. (2000). "The PSIPRED protein structure prediction server." Bioinformatics **16**(4): 404-405.

Meier, T., et al. (2007). "A tridecameric c ring of the adenosine triphosphate (ATP) synthase from the thermoalkaliphilic Bacillus sp. strain TA2.A1 facilitates ATP synthesis at low electrochemical proton potential." Mol Microbiol **65**(5): 1181-1192.

Meisinger, C., et al. (2001). "Protein import channel of the outer mitochondrial membrane: a highly stable Tom40-Tom22 core structure differentially interacts with preproteins, small tom proteins, and import receptors." Mol Cell Biol **21**(7): 2337-2348.

Meng, L., et al. (2007). "Actin binding and proline rich motifs of CR16 play redundant role in growth of vrp1Delta cells." Biochem Biophys Res Commun **357**(1): 289-294.

Milenkovic, D., et al. (2004). "Sam35 of the mitochondrial protein sorting and assembly machinery is a peripheral outer membrane protein essential for cell viability." J Biol Chem **279**(21): 22781-22785.

Mitchell, P. (1966). "Chemiosmotic coupling in oxidative and photosynthetic phosphorylation." Biol Rev Camb Philos Soc **41**(3): 445-502.

Miyatake, H., et al. (1999). "Dynamic light-scattering and preliminary crystallographic studies of the sensor domain of the haem-based oxygen sensor FixL from Rhizobium meliloti." Acta Crystallogr D Biol Crystallogr **55**(Pt 6): 1215-1218.

Mnaimneh, S., et al. (2004). "Exploration of essential gene functions via titratable promoter alleles." Cell **118**(1): 31-44.

Moczko, M., et al. (1992). "Identification of the mitochondrial receptor complex in Saccharomyces cerevisiae." FEBS Lett **310**(3): 265-268.

Model, K., et al. (2008). "Cryo-electron microscopy structure of a yeast mitochondrial preprotein translocase." J Mol Biol **383**(5): 1049-1057.

Model, K., et al. (2001). "Multistep assembly of the protein import channel of the mitochondrial outer membrane." Nat Struct Biol **8**: 361-370.

Model, K., et al. (2002). "Protein translocase of the outer mitochondrial membrane: role of import receptors in the structural organization of the TOM complex." J Mol Biol **316**(3): 657-666.

Mokranjac, D., et al. (2006). "Structure and function of Tim14 and Tim16, the J and J-like components of the mitochondrial protein import motor." Embo J **25**(19): 4675-4685.

Mokranjac, D. and W. Neupert (2009). "Thirty years of protein translocation into mitochondria: unexpectedly complex and still puzzling." Biochim Biophys Acta **1793**(1): 33-41.

Morgner, N., et al. (2006). "A new way to detect noncovalently bond complexes of biomolecules from liquid micro droplets by laser mass spectrometry." Aust. J. Chem. **59**: 109-114.
Morgner, N., et al. (2007). "A novel approach to analyze membrane proteins by laser mass spectrometry: from protein subunits to the integral complex." J Am Soc Mass Spectrom **18**(8): 1429-1438.
Morgner, N., et al. (2008). "Subunit mass fingerprinting of mitochondrial complex I." Biochim Biophys Acta **1777**(10): 1384-1391.
Myers, J. K., et al. (1995). "Denaturant m values and heat capacity changes: relation to changes in accessible surface areas of protein unfolding." Protein Sci **4**(10): 2138-2148.
Nakajo, K., et al. (2010). "Stoichiometry of the KCNQ1 - KCNE1 ion channel complex." Proc Natl Acad Sci U S A **107**(44): 18862-18867.
Naoe, M., et al. (2004). "Identification of Tim40 that mediates protein sorting to the mitochondrial intermembrane space." J Biol Chem **279**(46): 47815-47821.
Naveed, H., et al. (2009). "Predicting weakly stable regions, oligomerization state, and protein-protein interfaces in transmembrane domains of outer membrane proteins." Proc Natl Acad Sci U S A **106**(31): 12735-12740.
Neupert, W. (1997). "Protein import into mitochondria." Annu Rev Biochem **66**: 863-917.
Neupert, W. and J. M. Herrmann (2007). "Translocation of proteins into mitochondria." Annual Review of Biochemistry **76**: 6.1-6.27.
Nollert, P. (2002). "From test tube to plate: a simple procedure for the rapid preparation of microcrystallization experiments using the cubic phase method." J. Appl. Cryst. **35**: 637-640.
Oomen, C. J., et al. (2004). "Structure of the translocator domain of a bacterial autotransporter." Embo J **23**(6): 1257-1266.
Ott, M., et al. (2007). "The mitochondrial TOM complex is required for tBid/Bax-induced cytochrome c release." J Biol Chem **282**(38): 27633-27639.
Ott, M., et al. (2006). "Mba1, a membrane-associated ribosome receptor in mitochondria." Embo J **25**(8): 1603-1610.
Pace, C. N. (1986). "Determination and analysis of urea and guanidine hydrochloride denaturation curves." Methods Enzymol **131**: 266-280.
Palade, G. E. (1952). "The fine structure of mitochondria." Anat Rec **114**(3): 427-451.
Pauptit, R. A., et al. (1991). "Trigonal crystals of porin from Escherichia coli." J Mol Biol **218**(3): 505-507.
Perry, A. J., et al. (2006). "Convergent evolution of receptors for protein import into mitochondria." Curr Biol **16**(3): 221-229.
Perry, A. J., et al. (2008). "Structure, topology and function of the translocase of the outer membrane of mitochondria." Plant Physiol Biochem **46**(3): 265-274.
Popov-Celeketic, D., et al. (2008). "Active remodelling of the TIM23 complex during translocation of preproteins into mitochondria." Embo J **27**(10): 1469-1480.
Poynor, M., et al. (2008). "Dynamics of the preprotein translocation channel of the outer membrane of mitochondria." Biophys J **95**(3): 1511-1522.
Pratt, J. M., et al. (2006). "Multiplexed absolute quantification for proteomics using concatenated signature peptides encoded by QconCAT genes." Nat Protoc **1**: 1029-1043.
Preuss, M., et al. (2005). "Evolution of mitochondrial oxa proteins from bacterial YidC. Inherited and acquired functions of a conserved protein insertion machinery." J Biol Chem **280**(13): 13004-13011.
Proteau, A., et al. (2010). "Application of dynamic light scattering in protein crystallization." Curr Protoc Protein Sci **Chapter 17**: Unit 17 10.
Pusnik, M., et al. (2009). "The single mitochondrial porin of Trypanosoma brucei is the main metabolite transporter in the outer mitochondrial membrane." Mol Biol Evol **26**(3): 671-680.
Ramage, L., et al. (1993). "Functional cooperation of mitochondrial protein import receptors in yeast." EMBO J **12**(11): 4115-4123.
Rapaport, D., et al. (1998). "Dynamics of the TOM complex of mitochondria during binding and translocation of preproteins." Mol Cell Biol **18**(9): 5256-5262.

Rapaport, D., et al. (1997). "Mitochondrial protein import. Tom40 plays a major role in targeting and translocation of preproteins by forming a specific binding site for the presequence." J Biol Chem **272**(30): 18725-18731.

Rapaport, D., et al. (2001). "Structural requirements of Tom40 for assembly into preexisting TOM complexes of mitochondria." Mol Biol Cell **12**(5): 1189-1198.

Richardson, J. S. and D. C. Richardson (1988). "Amino acid preferences for specific locations at the ends of alpha helices." Science **240**(4859): 1648-1652.

Rimmer, K. A., et al. (2011). "Recognition of mitochondrial targeting sequences by the import receptors Tom20 and Tom22." J Mol Biol **405**(3): 804-818.

Robert, V., et al. (2006). "Assembly factor Omp85 recognizes its outer membrane protein substrates by a species-specific C-terminal motif." PLoS Biol **4**(11): e377.

Romero-Ruiz, M., et al. (2010). "Interactions of mitochondrial presequence peptides with the mitochondrial outer membrane preprotein translocase TOM." Biophys. J. **99**: in press.

Rostovtseva, T. K. and S. M. Bezrukov (1998). "ATP transport through a single mitochondrial channel, VDAC, studied by current fluctuation analysis." Biophys J **74**(5): 2365-2373.

Ryan, M. T. (2004). "Chaperones: inserting beta barrels into membranes." Curr Biol **14**(5): R207-209.

Saeki, K., et al. (2000). "Identification of mammalian TOM22 as a subunit of the preprotein translocase of the mitochondrial outer membrane." J Biol Chem **275**(41): 31996-32002.

Sagan, L. (1967). "On the origin of mitosing cells." J Theor Biol **14**(3): 255-274.

Salem, M., et al. (2010). "Revisiting glutaraldehyde cross-linking: the case of the Arg-Lys intermolecular doublet." Acta Crystallogr Sect F Struct Biol Cryst Commun **66**(Pt 3): 225-228.

Sass, E., et al. (2003). "Folding of fumarase during mitochondrial import determines its dual targeting in yeast." J Biol Chem **278**(46): 45109-45116.

Schägger, H. and G. von Jagow (1987). "Tricine-sodium dodecyl sulfate-polyacrylamide gel electrophoresis for the separation of proteins in the range from 1 to 100 kDa." Anal Biochem **166**(2): 368-379.

Scherer, P. E., et al. (1990). "A precursor protein partly translocated into yeast mitochondria is bound to a 70 kd mitochondrial stress protein." Embo J **9**(13): 4315-4322.

Schirmer, T., et al. (1995). "Structural basis for sugar translocation through maltoporin channels at 3.1 A resolution." Science **267**(5197): 512-514.

Schleiff, E. and T. Becker (2010). "Common ground for protein translocation: access control for mitochondria and chloroplasts." Nat Rev Mol Cell Biol.

Schmidt, O., et al. (2011). "Regulation of Mitochondrial Protein Import by Cytosolic Kinases." Cell.

Schmitt, S., et al. (2005). "Role of Tom5 in maintaining the structural stability of the TOM complex of mitochondria." J Biol Chem **280**(15): 14499-14506.

Schmitt, S., et al. (2006). "Proteome analysis of mitochondrial outer membrane from Neurospora crassa." Proteomics **6**(1): 72-80.

Schwartz, M. P. and A. Matouschek (1999). "The dimensions of the protein import channels in the outer and inner mitochondrial membranes." Proc Natl Acad Sci U S A **96**(23): 13086-13090.

Sebald, W., et al. (1979). "Preparation of Neurospora crassa mitochondria." Methods Enzymol **55**: 144-148.

Sheehan, D. (2009). Physical biochemistry: principles and applications Oxford, UK, John Wiley and Sons Ltd.

Sherman, E. L., et al. (2005). "Functions of the small proteins in the TOM complex of Neurospora crasssa." Mol Biol Cell **16**(9): 4172-4182.

Sideris, D. P., et al. (2009). "A novel intermembrane space-targeting signal docks cysteines onto Mia40 during mitochondrial oxidative folding." J Cell Biol **187**(7): 1007-1022.

Sirrenberg, C., et al. (1996). "Import of carrier proteins into the mitochondrial inner membrane mediated by Tim22." Nature **384**(6609): 582-585.

Sirrenberg, C., et al. (1997). "Functional cooperation and stoichiometry of protein translocases of the outer and inner membranes of mitochondria." J Biol Chem **272**(47): 29963-29966.

Söding, J. and A. N. Lupas (2003). "More than the sum of their parts: on the evolution of proteins from peptides." Bioessays **25**(9): 837-846.

Sokolova, L., et al. (2010). "Laser-induced liquid bead ion desorption-MS of protein complexes from blue-native gels, a sensitive top-down proteomic approach." Proteomics **10**(7): 1401-1407.
Söllner, T., et al. (1989). "MOM19, an import receptor for mitochondrial precursor proteins." Cell **59**(6): 1061-1070.
Söllner, T., et al. (1990). "A mitochondrial import receptor for the ADP/ATP carrier." Cell **62**(1): 107-115.
Spyropoulos, I. C., et al. (2004). "TMRPres2D: high quality visual representation of transmembrane protein models." Bioinformatics **20**(17): 3258-3260.
Sreerama, N. and R. W. Woody (2000). "Estimation of protein secondary structure from circular dichroism spectra: comparison of CONTIN, SELCON, and CDSSTR methods with an expanded reference set." Anal Biochem **287**(2): 252-260.
Sreerama, N. and R. W. Woody (2003). "Structural composition of betaI- and betaII-proteins." Protein Sci **12**(2): 384-388.
Sreerama, N. and R. W. Woody (2004). "Computation and analysis of protein circular dichroism spectra." Methods Enzymol **383**: 318-351.
Stan, T., et al. (2000). "Recognition of preproteins by the isolated TOM complex of mitochondria." EMBO J. **19**: 4895-4902.
Stutz, S. (2009). "Identifizierung und Charakterisierung der Proteintranslokase der mitochondrialen Außenmembran aus Bos taurus L." http://www.dr.hut-verlag.de/(ISBN 978-3-86853-031-5).
Suzuki, H., et al. (2004). "Membrane-embedded C-terminal segment of rat mitochondrial TOM40 constitutes protein-conducting pore with enriched beta-structure." J Biol Chem **279**(48): 50619-50629.
Suzuki, H., et al. (2000). "Characterization of rat TOM40, a central component of the preprotein translocase of the mitochondrial outer membrane." J Biol Chem **275**(48): 37930-37936.
Szabo, I. and M. Zoratti (1993). "The mitochondrial permeability transition pore may comprise VDAC molecules. I. Binary structure and voltage dependence of the pore." FEBS Lett **330**(2): 201-205.
Tal, M., et al. (1990). "A new method for stoichiometric analysis of proteins in complex mixture--reevaluation of the stoichiometry of E. coli ribosomal proteins." J Biochem Biophys Methods **21**(3): 247-266.
Timmis, J. N., et al. (2004). "Endosymbiotic gene transfer: organelle genomes forge eukaryotic chromosomes." Nat Rev Genet **5**(2): 123-135.
Tsaousis, A. D., et al. (2011). "A functional Tom70 in the human parasite Blastocystis sp.: implications for the evolution of the mitochondrial import apparatus." Mol Biol Evol **28**(1): 781-791.
Ujwal, R., et al. (2008). "The crystal structure of mouse VDAC1 at 2.3 A resolution reveals mechanistic insights into metabolite gating." Proc Natl Acad Sci U S A **105**(46): 17742-17747.
van der Laan, M., et al. (2006). "A role for Tim21 in membrane-potential-dependent preprotein sorting in mitochondria." Curr Biol **16**(22): 2271-2276.
van Wilpe, S., et al. (1999). "Tom22 is a multifunctional organizer of the mitochondrial preprotein translocase." Nature **401**(6752): 485-489.
Vergnolle, M. A., et al. (2005). "A cryptic matrix targeting signal of the yeast ADP/ATP carrier normally inserted by the TIM22 complex is recognized by the TIM23 machinery." Biochem J **385**(Pt 1): 173-180.
Vestweber, D., et al. (1989). "A 42K outer-membrane protein is a component of the yeast mitochondrial protein import site." Nature **341**(6239): 205-209 issn: 0028-0836.
Vial, S., et al. (2002). "Assembly of Tim9 and Tim10 into a functional chaperone." J. Biol. Chem. **277**: 36100-36108.
Voet, R. L. (1994). "Classification of vulvar dystrophies and premalignant squamous lesions." J Cutan Pathol **21**(1): 86-90.
Wagner, K., et al. (2008). "The assembly pathway of the mitochondrial carrier translocase involves four preprotein translocases." Mol Cell Biol **28**(13): 4251-4260.
Walther, D. M., et al. (2009). "Biogenesis of mitochondrial outer membrane proteins." Biochim Biophys Acta **1793**(1): 42-51.

Webb, C. T., et al. (2006). "Crystal structure of the mitochondrial chaperone TIM9 center dot 10 reveals a six-bladed alpha-propeller." Molecular Cell **21**(1): 123-133.
Weiss, M. S., et al. (1990). "The three-dimensional structure of porin from Rhodobacter capsulatus at 3 A resolution." FEBS Lett **267**(2): 268-272.
Welling-Wester, S., et al. (1988). "Effect of detergents on the structure of integral membrane proteins of Sendai virus studied with size-exclusion high-performance liquid chromatography and monoclonal antibodies." J Chromatogr **443**: 255-266.
Werhahn, W. and H. P. Braun (2002). "Biochemical dissection of the mitochondrial proteome from Arabidopsis thaliana by three-dimensional gel electrophoresis." Electrophoresis **23**(4): 640-646.
Werhahn, W., et al. (2003). "Identification of novel subunits of the TOM complex from Arabidopsis thaliana." Plant Physiology and Biochemistry **41**(5): 407-416.
Werhahn, W., et al. (2001). "Purification and characterization of the preprotein translocase of the outer mitochondrial membrane from Arabidopsis. Identification of multiple forms of TOM20." Plant Physiol **125**(2): 943-954.
Wiedemann, N., et al. (2003). "Machinery for protein sorting and assembly in the mitochondrial outer membrane." Nature **424**(6948): 565-571.
Wiley, W. C. and H. McLaren (1955). "Time-of-Flight Mass Spectrometer with Improved Resolution." The Review of Scientific Instruments **26**(12).
Williamson, M. P. (1994). "The structure and function of proline-rich regions in proteins." Biochem J **297 ( Pt 2)**: 249-260.
Winterfeld, S., et al. (2009). "Substrate-induced conformational change of the Escherichia coli membrane insertase YidC." Biochemistry **48**(28): 6684-6691.
Wojtkowska, M., et al. (2005). "An inception report on the TOM complex of the Amoeba Acanthamoeba castellanii, a simple model protozoan in mitochondria studies." J Bioenerg Biomembr **37**(4): 261-268.
Wu, Y. and B. Sha (2006). "Crystal structure of yeast mitochondrial outer membrane translocon member Tom70p." Nat Struct Mol Biol **13**(7): 589-593.
Yamamoto, H., et al. (2011). "Dual role of the receptor Tom20 in specificity and efficiency of protein import into mitochondria." Proc Natl Acad Sci U S A **108**(1): 91-96.
Yano, M., et al. (2004). "Mitochondrial import receptors Tom20 and Tom22 have chaperone-like activity." J Biol Chem **279**(11): 10808-10813.
Young, J. C., et al. (2003). "Molecular chaperones Hsp90 and Hsp70 deliver preproteins to the mitochondrial import receptor Tom70." Cell **112**(1): 41-50.
Zeth, K. (2010). "Structure and evolution of mitochondrial outer membrane proteins of beta-barrel topology." Biochim Biophys Acta.
Zeth, K., et al. (2000). "Crystal structure of Omp32, the anion-selective porin from Comamonas acidovorans, in complex with a periplasmic peptide at 2.1 A resolution." Structure **8**(9): 981-992.
Zeth, K. and M. Thein (2010). "Porins in prokaryotes and eukaryotes: common themes and variations." Biochem J **431**(1): 13-22.

# 6 Appendix

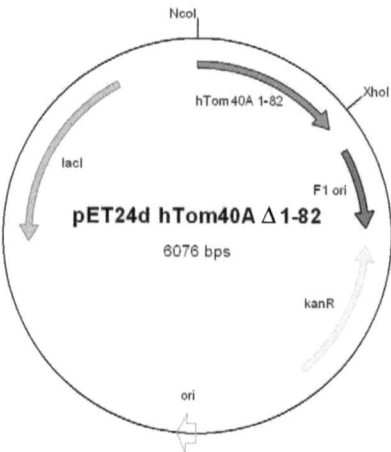

Figure 6.1: Plasmid pET24d with gene coding for the truncated protein hTom40AΔ1-82. Cloning sites NcoI and XhoI are indicated. Sequence of hTom40AΔ1-82 with hexa-histidinyl-tag. Weak amino acids, which have been mutated, are underlined.

```
  1  atggatttcg aggagtgcca ccggaagtgc aaggagctgt ttcccattca gatggagggt gtcaagctca
        m  d  f    e  e  c    h  r  k  c    k  e  l  f    p  i   q  m  e  g    v  k  l
 71  cagtcaacaa agggttgagt aaccattttc aggtcaacca cacagtagcc ctcagcacaa tcggggagtc
        t  v  n    k  g  l  s    n  h  f    q  v  n     h  t  v  a    l  s  t     i  g  e
141  caactaccac ttcggggtca catatgtggg gacaaagcag ctgagtccca cagaggcgtt ccctgtactg
        s  n  y  h    f  g  v    t  y  v     g  t  k  q    l  s  p    t  e  a      f  p  v  l
211  gtgggtgaca tggacaacag tggcagtctc aacgctcagg tcattcacca gctgggcccc ggtctcaggt
        v  g  d    m  d  n    s  g  s  l    n  a  q    v  i  h     q  l  g  p     g  l  r
281  ccaagatggc catccagacc cagcagtcga agtttgtgaa ctggcaggtg gacggggagt atcggggctc
        s  k  m    a  i  q  t    q  q  s    k  f  v    n  w  q  v     d  g  e     y  r  g
351  tgacttcaca gcagccgtca ccctgggaa cccagacgtc ctcgtgggtt caggaatcct cgtagccca
        s  d  f  t    a  a  v    t  l  g    n  p  d  v    l  v  g    s  g  i    l  v  a  h
421  tacctccaga gcatcacgcc ttgcctggcc ctgggtggag agctggtcta ccaccggcgg cctggagagg
        y  l  q    s  i  t    p  c  l  a    l  g  g    e  l  v    y  h  r  r     p  g  e
491  agggcactgt catgtctcta gctgggaaat acacattgaa caactggttg gcaacggtaa cgttgggcca
        e  g  t    v  m  s  l    a  g  k    y  t  l    n  n  w  l    a  t  v    t  l  g
561  ggcgggcatg cacgcaacat actaccacaa agccagtgac cagctgcagg tgggtgtgga gtttgaggcc
        q  a  g  m    h  a  t    y  y  h    k  a  s  d    q  l  q    v  g  v    e  f  e  a
631  agcacaagga tgcaggacac cagcgtctcc ttcgggtacc agctggacct gccaaggcc aacctcctct
        s  t  r    m  q  d    t  s  v  s    f  g  y    q  l  d     l  p  k  a    n  l  l
701  tcaaaggctc tgtggatagc aactggatcg tgggtgccac gctggagaag aagctcccac ccctgcccct
        f  k  g    s  v  d  s    n  w  i    v  g  a    t  l  e  k    k  l  p     p  l  p
771  gacactggcc cttggggcct tcctgaatca ccgcaagaac aagtttcagt gtggctttgg cctcaccatc
        l  t  l  a    l  g  a    f  l  n    h  r  k  n    k  f  q    c  g  f    g  l  t  i
841  ggcctcgagc accaccacca ccaccac
        g  l  e    h  h  h  h  h  h
```

128

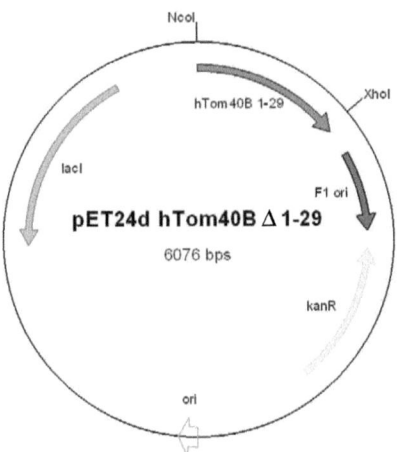

Figure 6.2: Plasmid pET24d with gene coding for the truncated protein hTom40BΔ1-29. Cloning sites NcoI and XhoI are indicated. Sequence of hTom40BΔ1-29 with hexa-histidinyl-tag.

```
  1  atggatttcg atgagctgca ccgtctatgc aaagatgtat tcccagcaca gatggaggga gtgaagctcg
     m   d   f   d   e   l   h   r   l   c   k   d   v   f   p   a   q   m   e   g   v   k   l
 71  ttgtcaacaa ggttctgagc agccatttcc aggtggcgca cactatacac atgagtgccc tgggcttgcc
     v   v   n   k   v   l   s   s   h   f   q   v   a   h   t   i   h   m   s   a   l   g   l
141  gggatatcac ctccatgcgg cctatgcagg ggattggcag ctcagtccca ctgaggtgtt ccccactgtg
     p   g   y   h   l   h   a   a   y   a   g   d   w   q   l   s   p   t   e   v   f   p   t   v
211  gtagggggata tggacagcag tggcagcctg aacgcccagg tcttgctcct cttggcagag cggctccgag
     v   g   d   m   d   s   s   g   s   l   n   a   q   v   l   l   l   a   e   r   l   r
281  ctaaggctgt cttccagacg cagcaggcca agttcctgac atggcagttt gatggcgagt atcggggaga
     a   k   a   v   f   q   t   q   q   a   k   f   l   t   w   q   f   d   g   e   y   r   g
351  tgactacaca gccactctga ccctaggaaa tcctgacctg attggggagt cggtgatcat ggttgctcac
     d   d   y   t   a   t   l   t   l   g   n   p   d   l   i   g   e   s   v   i   m   v   a   h
421  ttcctgcaga gcctcactca tcggctggtg ctgggaggag agctagttta tcaccggcgg ccaggcgaag
     f   l   q   s   l   t   h   r   l   v   l   g   g   e   l   v   y   h   r   r   p   g   e
491  aggggggccat cttgacactg gctgggaagt actcggctgt acactgggta gctacattga atgtgggatc
     e   g   a   i   l   t   l   a   g   k   y   s   a   v   h   w   v   a   t   l   n   v   g
561  aggcggggcc catgcaagtt actaccacag ggcaaatgaa caggttcagg ttggagtgga gtttgaggca
     s   g   g   a   h   a   s   y   y   h   r   a   n   e   q   v   q   v   g   v   e   f   e   a
631  aacacaaggc tacaagacac aacattctcc tttggttacc acctgactct gccccaggcc aacatggtat
     n   t   r   l   q   d   t   t   f   s   f   g   y   h   l   t   l   p   q   a   n   m   v
701  ttagaggctt ggtggatagt aactggtgtg taggtgctgt gctggagaag aagatgcccc ctctgcctgt
     f   r   g   l   v   d   s   n   w   c   v   g   a   v   l   e   k   k   m   p   p   l   p
771  cacccctagcc cttggagcct tcctcaatca ctggcgcaac agattccatt gtggcttcag catcactgtg
     v   t   l   a   l   g   a   f   l   n   h   w   r   n   r   f   h   c   g   f   s   i   t   v
841  ggcctcgagc accaccacca ccaccac
     g   l   e   h   h   h   h   h   h
```

Die VDM Verlagsservicegesellschaft sucht für wissenschaftliche Verlage abgeschlossene und herausragende

## Dissertationen, Habilitationen, Diplomarbeiten, Master Theses, Magisterarbeiten usw.

für die kostenlose Publikation als Fachbuch.

Sie verfügen über eine Arbeit, die hohen inhaltlichen und formalen Ansprüchen genügt, und haben Interesse an einer honorarvergüteten Publikation?

Dann senden Sie bitte erste Informationen über sich und Ihre Arbeit per Email an *info@vdm-vsg.de*.

**Sie erhalten kurzfristig unser Feedback!**

VDM Verlagsservicegesellschaft mbH
Dudweiler Landstr. 99          Telefon +49 681 3720 174
D - 66123 Saarbrücken          Fax     +49 681 3720 1749

**www.vdm-vsg.de**

Die VDM Verlagsservicegesellschaft mbH vertritt

Printed by Books on Demand GmbH, Norderstedt / Germany